KT-484-925

British Isles

A Natural History

British

Isles
A Natural History

Alan Titchmarsh

This book is published to accompany
British Isles – A Natural History, first broadcast on BBC1 in 2004
Executive Producer: Mike Gunton
Supervisory Producer: Michael Bright

First published in 2004
Reprinted 2004

Published by BBC Books, BBC Worldwide Ltd,
Woodlands, 80 Wood Lane, London W12 0TT

ISBN 0 563 52162 7

Commissioning Editor: Nicky Ross
Project Editor: Martin Redfern
Copy Editor: Ben Morgan
Designer: Andrew Barron @ Thextension
Art Director: Linda Blakemore
Picture Researcher: Kath Kollberg
Production Controller: Christopher Tinker

Alan Titchmarsh would like to thank the following people for their invaluable help in the writing of this book: chapter 1, Stuart Armstrong; chapter 2, Dan Tapster; chapter 3, Ian Gray; chapter 4, Chris Cole; chapter 5, Charlotte Scott and Venetia Scott; chapter 6, Jessica Pailthorpe; chapter 7, Dan Tapster; Places to Visit, Mary Colwell; and to Patrick Morris for his editorial input.

Set in Foundry Sans and Veljovic
Colour origination and printing by Butler & Tanner Ltd

Front cover picture captions
Stonehenge, Wiltshire; Sycamore Gap by Hadrian's Wall;
Alan on the summit of Aonach Mor, Scotland.

Back cover picture captions
Combine harvester at work; The Seven Sisters, Sussex;
Looking towards Ullswater in the Lake District.

Introduction

I am unashamed to boast that I love being British, and I'm happy to confess that I could not live anywhere else. It's a feeling that is not born of xenophobia or small-mindedness, but comes simply from a love of the landscape in which I have grown up, lived and worked for more than half a century. Unlike Nancy Mitford's father, I do not dislike 'abroad', neither do I distrust foreigners. But, more than anything, I love coming home.

I was born and brought up in the Yorkshire Dales, and holidayed during my childhood in the Lake District and Scotland (not forgetting Blackpool). I now live in Hampshire and the Isle of Wight, and often spend my holidays in Cornwall. This, coupled with 25 years of working in television and travelling the length and breadth of the country for programmes as varied as **Gardeners' World** and **Songs of Praise**, have served only to increase my love of the British Isles and their astonishing countryside.

There are blots on the landscape, true enough, but within a stone's throw of all our houses – whether in city or village – there is some piece of landscape that is worth celebrating and exploring.

Making the television series of **British Isles – A Natural History** took two years, and during that time I enjoyed such diverse pursuits as walking Hadrian's Wall, watching swallowtail butterflies on the Norfolk Broads, red squirrels in Lancashire and seals in the Firth of Forth. I have rowed in a hailstorm on Ullswater, flown over the Hebrides (sideways) in an RAF jet fighter, been cut off by the tide at Hartland Quay in Devon, and stood level with the snow-capped peak of Ben Nevis in sparkling sunshine.

You do not need to journey to the Grand Canyon or to Ayers Rock to see breathtaking scenery. From the Scillies in the south to the northernmost Scottish coast, from Killarney in the west of Ireland, and across Wales to East Anglia, I have drunk my fill of the best views on

Earth and encountered rare and common animals and birds, wild flowers and fossils that would leave anyone with the slightest interest in natural history reeling with pleasure.

My training is as a gardener. But at the age of ten or eleven I became a member of the Wharfedale Naturalists' Society, and my love of nature has always gone hand in hand with my passion for growing plants. The two are, for me, inextricably linked. I still have a bird book by my desk, and the fact that I refuse to use garden chemicals is due entirely to the fact that I consider that other forms of life have every bit as much right to use my garden as I do.

If anything, the making of the series and the production of this book have confirmed that my love of these islands is not misplaced. They are uniquely beautiful and amazingly varied. We mock our weather without realizing that its very inconsistencies are responsible for the beauty of our surroundings.

If **British Isles – A Natural History** does anything to increase your fondness for the land that is our home, it will have done its job. Not simply in a finger-wagging way – though I earnestly hope that it will awaken in those who are complacent a need to take care of these unique islands – but also in telling a story that is packed with adventure and incident, thrilling tales and wonderful wildlife. How it and we came to be here justly ranks as one of the greatest stories ever told. With consideration and understanding, and a proprietorial sort of pride in the place, it is one that can continue to unfold.

Alan Titchmarsh

Hampshire, 2004

Solid Foundations

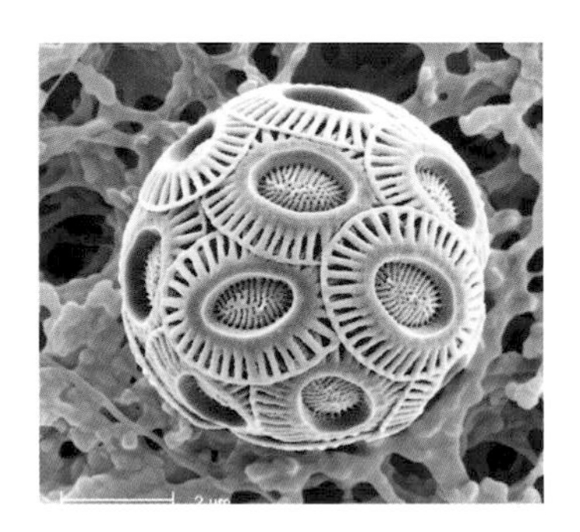

1

3 billion to 3 million years ago

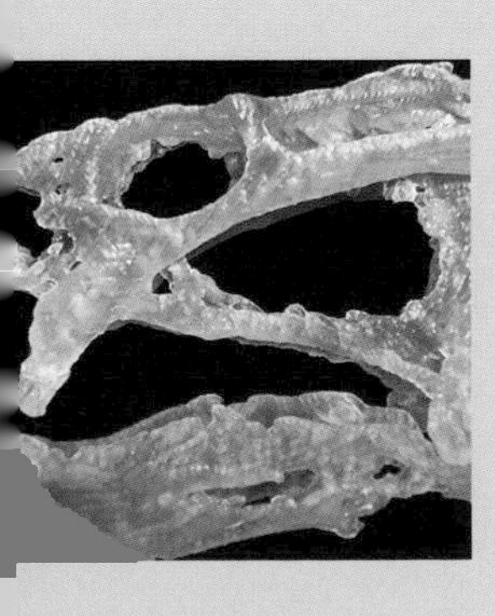

IN OUR MIND'S EYE, MOST OF US HAVE A PICTURE OF THE BRITISH ISLES. SOME BITS OF IT ARE CRYSTAL CLEAR – THE PART OF THE COUNTRY IN WHICH WE LIVE OR WERE BORN – AND OTHER BITS WILL BE NO MORE THAN VAGUE IMPRESSIONS. THE CLASSROOM SHAPED MY OWN VIEW OF THE COUNTRY. I WAS BORN AND BROUGHT UP IN YORKSHIRE, AND THE COAST OF THAT COUNTY WILL FOREVER REMIND ME OF THE FACE OF A BEARDED MAN. WALES ALWAYS SEEMED TO ME TO LOOK LIKE THE HEAD OF A PIG – NO INSULT

INTENDED TO ITS INHABITANTS. WHATEVER OUR IMPRESSIONS, TO US THE BRITISH ISLES HAVE ALWAYS BEEN THAT CLUSTER OF ODDLY SHAPED ISLANDS SITTING QUIETLY OFF THE WESTERN EDGE OF EUROPE. BUT TRAVEL BACK IN TIME AND THE OUTLINE OF THE ISLANDS WE KNOW TODAY WOULD HAVE BEEN UNRECOGNIZABLE. IT'S TAKEN BILLIONS OF YEARS FOR THE LANDSCAPE TO REACH ITS PRESENT SHAPE AND CHARACTER. EVERY PART OF IT, FROM THE CRAGGY MOUNTAINS OF SCOTLAND TO THE ROLLING CHALK DOWNS OF SOUTHERN ENGLAND, IS THE PRODUCT OF AN EXTRAORDINARY JOURNEY THAT HAS SEEN BRITAIN AND IRELAND TRAVEL FROM THE POLAR SEAS OF THE ANTARCTIC, ACROSS THE WARM TROPICS, AND ON TO THE TEMPERATE LATITUDES IN WHICH WE FIND OURSELVES TODAY.

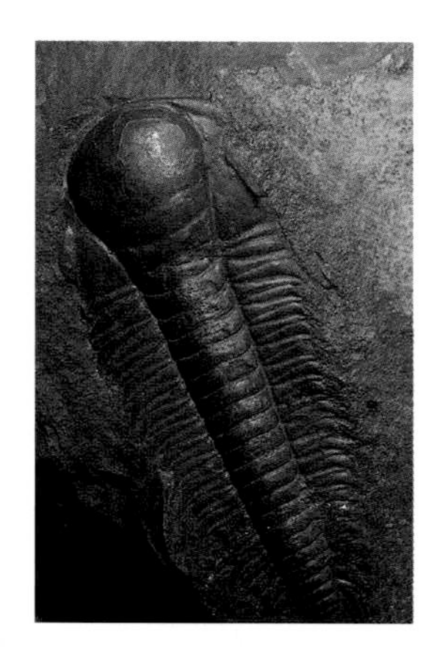

The clues to this journey are to be found under our feet in the Earth's foundations – the very rocks of our islands. Each piece of evidence helps us put together a picture of an astonishing journey. It is difficult to imagine that, at various times, the British Isles were desert, rainforest, the floor of a tropical sea, or thrust skywards into mountains of Himalayan proportions. But they were. The journey began 3 billion years ago, and it is far from over.

How Old Are the British Isles?

Just as most people have a birth certificate, so too does the landscape itself. The British Isles' birth certificates are permanently etched into their foundations (*see box* 'Birth Certificates', p. 15). Rocks might not seem to be the most informative of objects – and at school, geology seemed as dry as dust to me – but every type of rock that makes up these islands has kept details of its history and origins. By piecing together these clues, we can discover postcards from the past, and they make fascinating reading. The variety of rocks that makes up the British Isles is one of the most complex in the world, including an almost complete sequence of different rock types from all the prehistoric eras. Each rock type records a different leg of a remarkable journey that carried these islands around the globe. And the oldest certificate is to be found in one of the most remote corners of Britain.

In far northwest Scotland, the Outer Hebridean island of Lewis is a place of peat moors and rounded hills. It looks and feels ancient. The hills give the impression of hunkering down, backs to the wind in a vain attempt to avoid the cruel weather thrown in from the Atlantic. The rocks of this island are impervious, draining very poorly and producing a landscape of lakes and rain-sodden bogs; the area includes some of the last remaining peat bogs in the country.

(ABOVE) THE PENNINES ARE FORMED FROM A SEDIMENTARY ROCK CALLED MILLSTONE GRIT, NAMED AFTER THE MILLSTONES THAT WERE ONCE CARVED FROM THIS ROCK.

(OPPOSITE) TRILOBITE FOSSILS HIDE SOME REMARKABLE SECRETS TO OUR ISLANDS' HISTORY. THEY TELL US THAT WE WERE ONCE VERY MUCH A DIVIDED NATION.

(PREVIOUS PAGE) THE STANDING STONES OF CALLANISH MARK THE BIRTH OF BRITAIN, BOTH HISTORICALLY AND GEOLOGICALLY.

For centuries, locals have been digging the peat to use as fuel. In the mid-nineteenth century, diggers in the west of the island unearthed something far more exciting than peat: the Standing Stones of Callanish, one of the oldest prehistoric sites in Britain. Erected more than 6000 years ago, this stone circle was part ceremonial site and part celestial calendar, mapping the passage of time for the ancient communities that lived on Lewis. Little did they know, the ancient islanders had built their stone circle from the rock that marks the very birth of Britain: Lewisian gneiss (pronounced 'nice'). Geologists have dated this rock as nearly 3 billion years old, making it not just the oldest rock in Britain but one of the oldest types of rock in the world. This rock is visible at the surface only in northwest Scotland, but it potentially underlies the whole of the British Isles. It is our basement, our foundation, and its formation 3 billion years ago marks the birth of our islands.

Geologists classify the bewildering variety of rocks that exist into just three main groups, depending on the way they form.

The 'igneous' (fire-formed) rocks, such as granite and basalt, form when molten rock cools and solidifies. Granite forms deep under ground when 'magma' (scalding, molten rock that is thicker than treacle) solidifies slowly, turning into crystals. Basalt forms more suddenly, from volcanic lava that freezes as it bursts through the Earth's crust. Parts of Snowdonia, the Lake District, Dartmoor, the Scilly Isles, the Cairngorms and even the centre of Edinburgh (Castle Rock) are the remnants of ancient eruptions that brought huge quantities of magma onto or close to the surface.

Once a rock is exposed at the Earth's surface, it begins to crumble. Water, wind, ice and fluctuating temperatures all attack it, chiselling off fragment after fragment. The tiny crumbs are blown away and washed downstream by rivers, where they eventually settle. Only tiny amounts wear off solid rock each day, but these add up over millions of years to huge quantities. Where the debris settles, layer after layer of crumbs build up. The deeper this sediment becomes, the greater the pressure on the lower layers, which are gradually compressed to form new rock. This is 'sedimentary' rock, such as sandstone, limestone and shale. These rocks are where we find fossils, which give us a window into the lost world of extinct animals and plants that once made Britain their home.

The third way that rock can form involves a bit of remodelling. The 'metamorphic' rocks form when other types of rock are subjected to huge pressures and temperatures deep under ground, without melting the rock outright. The pressure and heat slowly cook the rock and turn it into new forms, such as our most ancient rock, Lewisian gneiss, and one of our most well known: slate.

(LEFT) MANY OF BRITAIN'S WILDEST PLACES, INCLUDING THE LAKE DISTRICT, ARE THE REMNANTS OF VOLCANIC ERUPTIONS.

(FAR LEFT) THE MAJORITY OF FOSSILS COME FROM SEDIMENTARY ROCKS, INCLUDING THESE AMMONITES CAPTURED IN A PIECE OF SHALE.

(BELOW) MUCH OF BRITAIN IS KEPT DRY BY SLATE, A METAMORPHIC ROCK WHICH IS SIMPLY RE-FORMED MUDSTONE.

Lewisian gneiss is a metamorphic rock (*see box* 'Rock Types'), which means it was created when even older rocks were cooked and remodelled deep in the Earth's crust. It formed when the Earth was in its infancy – a violent, volcanic time. There was little or no oxygen, oceans were just beginning to form from the breath of volcanoes, and the land was littered with debris and solidified lava from eruptions.

The first 2 billion years – by far the majority of our islands' history – can be glimpsed only through the most fogged of geological spectacles. We can date the rocks, but that's about all. If there was life in the British Isles at that time, there are no records of it now – the rocks have been so buckled and twisted over the aeons that any fossils have long since disappeared.

The first sign of life in Britain comes from the Torridonian mountains on the northwest Scottish mainland, where sedimentary rocks about a billion years old bear fossils of the simplest bacteria – not the most exciting of creatures, but remnants from a time when life was just gaining a foothold on the first rung of the evolutionary ladder. Farther down the coast, on Islay, are signs that life was flourishing in the sea. Here, bacterial colonies became fossilized into mushroom-shaped stools called stromatolites, that now litter the foreshore by one of the island's famous whisky distilleries. (Whisky is the Gaelic word for 'waters of life', an apt expression for these shores 1 billion years ago.) And bacteria are all we can find in our geological record for the next 400 million years.

Birth Certificates

A rock is a collection of chemical compounds called minerals, which are usually sandwiched together as tiny crystalline grains. Like a birth certificate, the minerals in a rock can tell us both when and where that rock was born. One of the minerals that best tells us a rock's age is uranium. Uranium isn't confined to nuclear power stations; it's found in all rock types, though more in some than others.

A walk over Dartmoor or the Cairngorms, where the granite bedrock is rich in uranium, can get any Geiger counter to sing. Over time, uranium slowly 'decays' into lead, which is more stable. Since this breakdown occurs at a constant rate, the ratio of uranium to lead gives a measure of the rock's age.

The mineral magnetite, which is rich in magnetic iron, can help us work out where a rock was born. When igneous rock forms deep under ground as molten rock (magma) cools, particles of magnetite within it align themselves with the Earth's magnetic field. Once the rock has cooled, these tiny magnetite compasses remain frozen permanently in position. They record not only the direction of north and south, but the 'dip angle' of the magnetic field, and this reveals the latitude at which the rock formed.

United Kingdom?

Discovering the birthplace of rocks has given us a tremendous insight into our past. Around 520 million years ago, the United Kingdom was anything but united. Careful analysis of Scottish rocks of this age reveals that they formed about 30° south of the equator, at about the same latitude as South Africa. Yet rocks just a few hundred miles away in England formed 60° south of the equator – almost 3000 miles (5000 km) away from the Scottish rocks, and much closer to the South Pole. Difficult as it is to believe, England and Scotland were once as far apart as Britain and North America are today.

More evidence of this ancient separation comes from fossils. The largest fossils from around 520 million years ago are trilobites – sea creatures that look like giant woodlice. The trilobites' hard exoskeletons left behind fossils so detailed that different species can easily be recognized. Such fossils are found in both Scotland and England, but the species are completely different. The Scottish trilobites are never found in England, and the English ones never turn up in Scotland. Since trilobites lived in shallow coastal seas and not in deep water, a vast ocean must have separated England from Scotland.

There was one place where those Scottish species were found: Newfoundland. As we now know, Scotland and Northern Ireland were both once attached to the eastern part of Canada in an ancient continent called Laurentia, from which North America later formed. England, Wales and southern Ireland were part of another ancient continent, Avalonia, which lay much further south. Between the two continents was an ocean as large as the present-day Atlantic, called the Iapetus Ocean. How could countries that have been neighbours for so long have started so far apart? And, more importantly, how did the landmasses that make up Britain and Ireland finally unite? It's all a matter of plate tectonics (*see box* 'Plate Tectonics', pp. 18–19).

North Wales is pockmarked with ancient volcanic remains. From the rolling farmland of the Llyn Peninsula to the well-trampled Snowdon and the isolated beauty of Cader Idris, much of this land is made from volcanic ash and magma. But these eruptions weren't always on land – they were from submarine volcanoes. We know this because the ash that rained down from their eruptions trapped huge numbers of marine fossils called graptolites. You can even find these on the summit of Snowdon.

The Welsh volcanoes formed around 520 million years ago over a subduction zone – an area where two of the jigsaw pieces in the Earth's crust push violently together, causing volcanoes and earthquakes. This zone of activity ran along the coast of the ancient continent of Avalonia and also helped to create today's Lake District; just like Snowdon, the moody peaks of Scafell Pike and Helvellyn are the remains of extinct volcanoes. Subduction zones occur where the edge of one of the plates in the Earth's crust is forced under another and destroyed. In this case, the plate that made up the ocean floor was being pushed under the continent of Avalonia, much as the Pacific floor is being slowly pushed under the coast of California today. And this means that the ancient ocean was slowly shrinking. England was drifting north, shifting ever closer to Scotland, though it would take another 120 million years before they met.

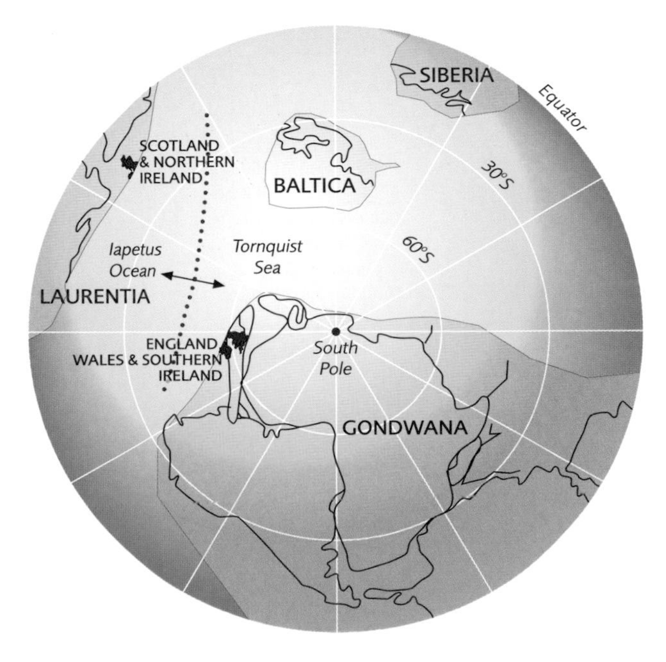

landmass
sea
ocean ridge
original position of present-day coastlines

(LEFT) THE GLOBE 520 MILLION YEARS AGO. AT THAT TIME SCOTLAND AND ENGLAND SAT ON DIFFERENT CONTINENTS, SEPARATED BY AN OCEAN THE SIZE OF THE MODERN ATLANTIC.

(OPPOSITE) SNOWDON IS A LONG-EXTINCT MARINE VOLCANO. EVEN AT THE SUMMIT YOU CAN FIND MARINE FOSSILS CALLED GRAPTOLITES IN ITS DARK VOLCANIC ROCK.

The Earth's crust is not a solid, unbroken layer of rock. It's a jigsaw of interlocking segments, called tectonic plates, that slowly drift about, driven by currents in the underlying mass of molten rock. Where these plates meet, huge forces are unleashed, causing earthquakes and volcanoes. Think of California, Japan, Turkey, Iran – all these countries sit at the edge of tectonic plates and therefore suffer some of the most violent earthquakes and volcanoes on the planet. Every newsflash reporting another quake or eruption is a reminder that the Earth's crust is forever shifting, albeit very slowly.

Tectonic plates are not just moving – they are continually being created and destroyed. In the 1960s, scientists made a ground-breaking discovery while investigating the mid-ocean ridge, a range of volcanic mountains that runs down the middle of the Atlantic Ocean. The scientists found that new crust is forming along the length of the ridge from lava pushing up from below. As the new crust forms, older crust is shunted away, causing the Atlantic Ocean to get about 0.4–4 in (1–10 cm) wider each year – the same rate as fingernails grow.

There is only a finite amount of space on the Earth's surface, so if new crust is forming in some places, old crust must be getting destroyed somewhere else. Japan and California both sit over areas where an oceanic plate and a continental plate are being pushed together. The continental plate, being lighter, is riding up on the oceanic plate. And the oceanic plate is being thrust down into what geologists call a

THE SEA FLOOR

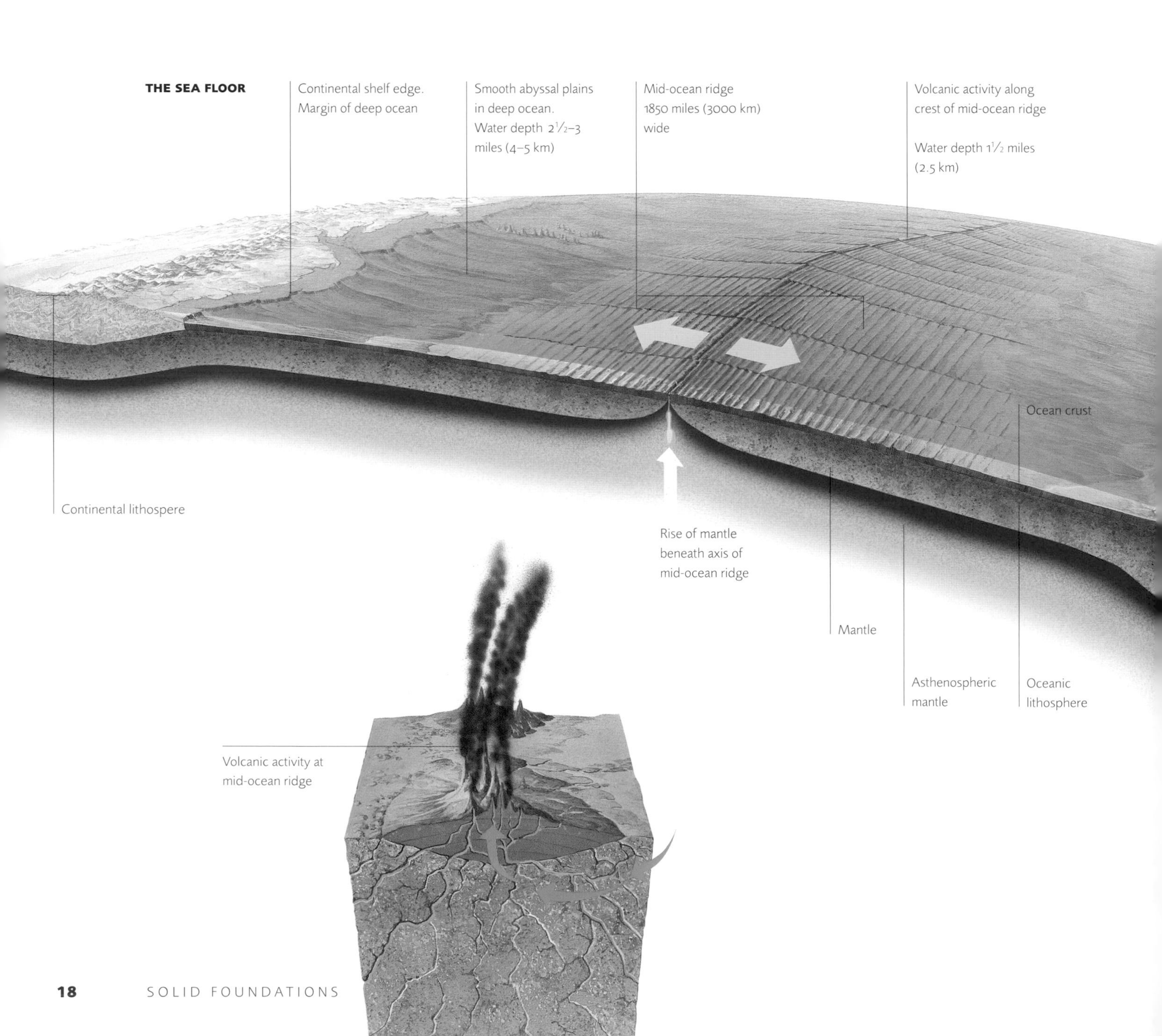

'subduction zone', where the crust melts and turns into magma, some of which squeezes back to the surface to form volcanoes.

Today Britain sits safely in the heart of the European tectonic plate, but 520 million years ago, England and Wales were right at the edge of a tectonic plate. The hills of Snowdonia and the Lake District are the remains of volcanoes from that time.

Subduction zones

Around the edge of some oceans, the whole strong outer part of the Earth, called the lithosphere or the plate, sinks as part of the process of subduction. Where the oceanic plate dives below the continental plate, a deep trench forms in the ocean floor. The ocean crust is being subducted beneath the continental plate, pushing up mountains in the process. Sedimentary rocks on the sea floor are scraped off and plastered onto the edge of the overlying plate, forming an accretionary prism. At depth, the sinking plate heats up, causing the crust to melt. The molten rock pools together and rises up towards the surface. Much of this molten rock solidifies beneath the surface, building up the crust. But some erupts in the great chains of volcanoes that form volcanic arcs above the subducting plates.

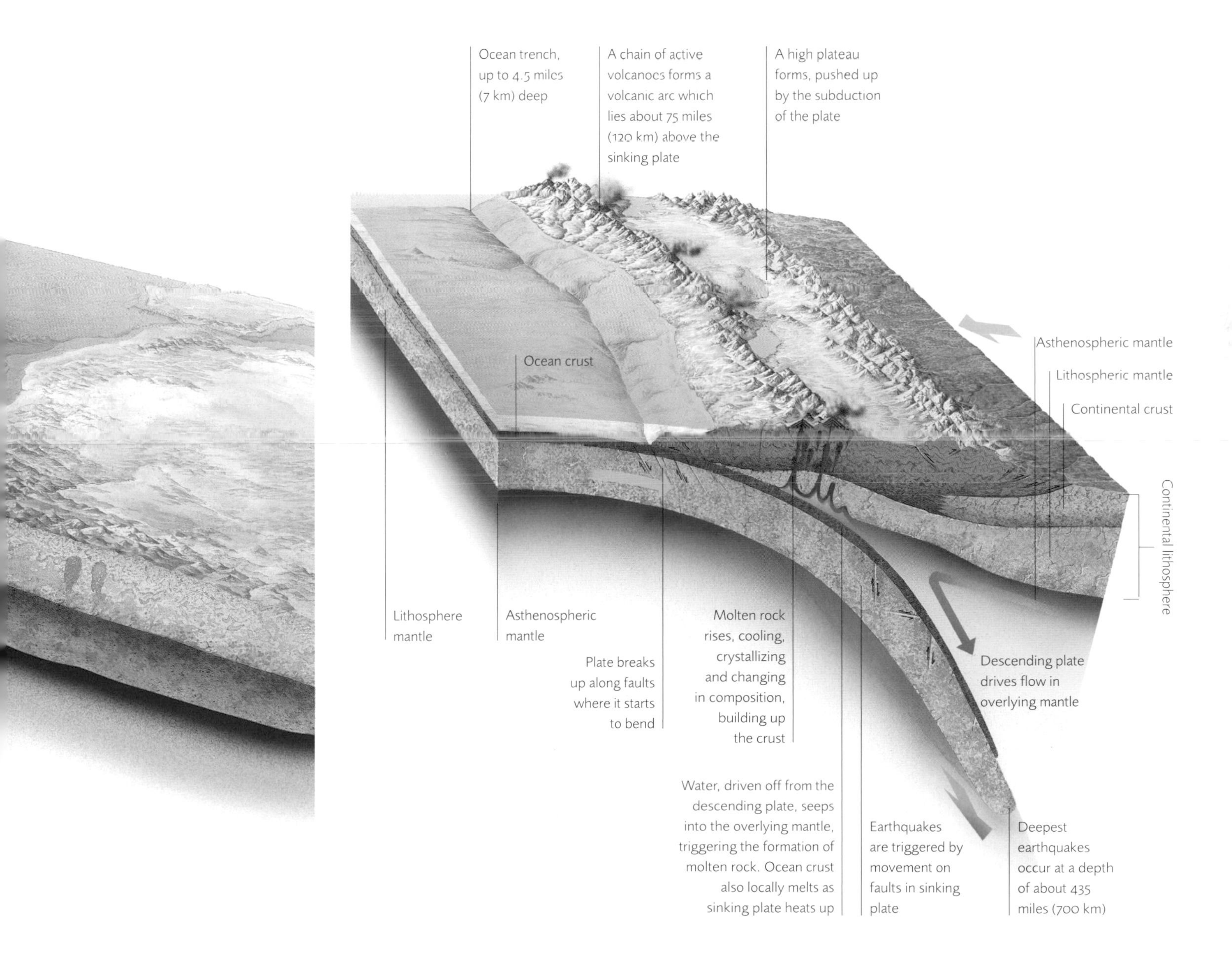

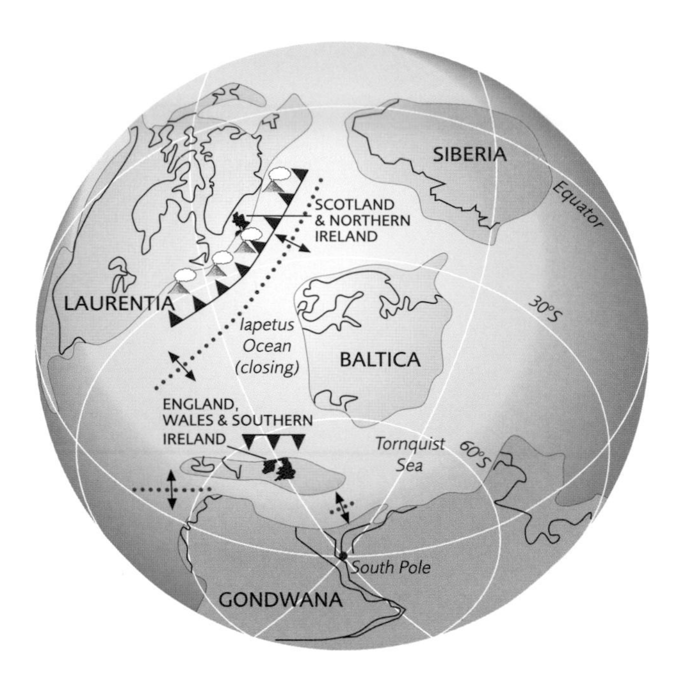

Our Own Himalayas

Around 400 million years ago, England, Ireland, Scotland and Wales were finally brought together as the Iapetus Ocean closed. It was a union that had a catastrophic effect on the landscape. As the continental plates collided, Scotland crumpled like a rug being pushed against a wall. Its crust buckled and bent, slowly thickening into a range of enormous mountains – the Caledonides – in just the same way that the Himalayas formed when India crashed into Eurasia. For a while, Scotland would have looked like Nepal.

The remnants of this epic event are found throughout the Highlands, where there are large numbers of parallel fault lines – vast cracks in the land where blocks of crust have slid past each other, driven by the force of two continents grinding together. The Great Glen – home to Loch Ness, our deepest lake – is the most famous. On a map it looks as though a knife has sliced through this part of Scotland, leaving a long, straight scar. But the greatest legacy of the continental collision are the weathered remains of the mighty Caledonides: the Grampian Mountains, which contain Britain's highest peaks, including Ben Nevis and the Cairngorms. Our wildest landscapes are to be found in the Grampians, which provide a refuge for some of our most celebrated and elusive animals, including golden eagles, red deer and pine martens.

Further south, between Glasgow and Carlisle, are the Scottish Uplands, which formed from the floor of the Iapetus Ocean as it was squeezed between colliding continents, forcing it up into hills. And to the south of this range is the join itself, the 'Iapetus suture', a stitch of twisted rock that linked Scotland and England for the very first time. This is not a join

(ABOVE LEFT AND RIGHT, AND OPPOSITE ABOVE LEFT) TECTONIC MAPS OF THE GLOBE 500, 460 AND 440 MILLION YEARS AGO RESPECTIVELY, SHOWING THE CLOSURE OF THE IAPETUS OCEAN. OVER 120 MILLION YEARS, THE IAPETUS CLOSED SLOWLY, AND ENGLAND, WALES, SCOTLAND AND IRELAND WERE UNITED FOR THE FIRST TIME.

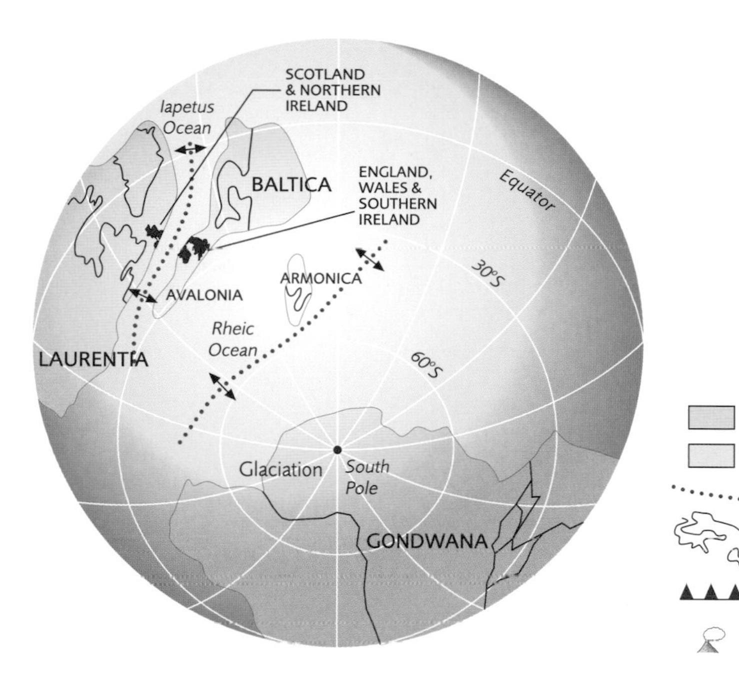

that you can see – it remains hidden many hundreds of feet below the moorland, visible only through the interpretations of science. By a twist of fate, it almost exactly follows the modern political boundary between the two countries.

Scotland's ancient mountains once rivalled the Himalayas in size, but today only a handful of peaks are more than 3000 feet (1000 m) high, and Ben Nevis, Britain's tallest mountain, is just 4409 feet (1344 m) tall – barely knee-high to a Himalayan summit. What monumental forces could have tamed the once-vast peaks of the Caledonides?

Any walker knows that you must wear the right clothing for a hike in the hills. Whatever the time of year, you risk getting drenched by a sudden shower. Hills and mountains are wet places, with notoriously changeable weather. They force air to rise as it flows over them, which makes water vapour cool, condense and fall as rain or snow. And water is surprisingly destructive. It seeps into crevices in rock, freezes and

(ABOVE RIGHT) BORN BY A CONTINENTAL COLLISION, THE HIGHLANDS OF SCOTLAND CONTAIN BRITAIN'S WILDEST LANDSCAPES AND ARE A REFUGE FOR SOME OF OUR MOST ELUSIVE ANIMALS, INCLUDING THE GOLDEN EAGLE.

(LEFT) SCOTLAND'S OLD RED SANDSTONE IS RICH IN OUR EARLIEST FISH FOSSILS, SUCH AS THIS PRIMITIVE SPECIES OF SHARK CALLED *COCCOSTEUS*.

splits off shards of rock as the ice expands. Rivers and streams eat their way into the ground, and glaciers carve away vast tracts of land. In time, these processes can raze mountains to the ground. Over the best part of 400 million years, the mighty peaks of the Caledonides have been ground down to the rounded stumps of granite that we see today.

The newly formed Britain was sitting just south of the equator around 400 million years ago. Conditions were desert-like, comparable with Arizona in the southwest USA. Monsoons would have regularly drenched the slopes of the Caledonides, sweeping away fragments of rock and speeding up the process of erosion. The debris was washed downstream by rivers and dumped in nearby seas and lakes, where it eventually became recycled into a new type of rock: Old Red Sandstone. In parts of Scotland, this sandstone forms layers up to 6 miles (10 km) thick. And trapped inside it are some fascinating relics from the past.

The ancient deserts of Britain would have been far more barren than any deserts today. There were no cacti or thorny shrubs, as plants were only just making the evolutionary leap from the sea and onto land. As for land animals, there weren't many more than harvestmen and mites. But the oceans and lakes were absolutely teeming with life, as the Old Red Sandstone reveals.

People have quarried Scotland's Old Red Sandstone for centuries to use it as a building material. At Achannaras in Caithness, near the northeastern tip of Scotland, quarry workers found more than just rock. They stumbled across hundreds of fossilized fish that had lain entombed for 380 million years. These were some of the oldest fish fossils ever discovered. What's more, they were freshwater fish (older fossil fish are all marine) – clear evidence that fish were beginning to conquer new environments.

It seems that Achannaras was once on the southern shore of an enormous lake (Lake Orcadia) that was packed with prehistoric fish. Most were tiny, but there were some larger predators, one of the most common being an extinct species of armoured shark called *Coccosteus*. This was a time when fish were without doubt the dominant predators on the planet, though there were other creatures to be wary of too, including a giant sea scorpion (*Pterygotus*) well over 7 feet (2 m) long. To me, those dull old rocks that seemed so lifeless at school have proved themselves fascinating treasure troves of our natural history.

To the north of Caithness are the Orkney Islands, which formed from the same chunk of Old Red Sandstone. The high cliffs and sea stacks that skirt the islands are home to huge numbers of seabirds, which arrive in April each year to nest on horizontal ledges in the rock. The ledges are the eroded remains of the floor of Lake Orcadia. The annual monsoons that drenched the Caledonides some 380 million years ago washed layer after layer of debris into the lake, building up the sandstone in bands. Today, these layers of sandstone give the Old Man of Hoy its stripy appearance and provide snug ledges in this seabird high-rise.

Old Red Sandstone is not confined to northeast Scotland – it crops up in Herefordshire, where it colours the russet soil, and it forms the Black Mountains and Brecon Beacons of southern Wales. There are also large areas of Old Red Sandstone in Devon, which gives its name to the geological era in which the rock formed: the Devonian. Devon's sandstone is not actually red but black, a sign that it formed in a deep sea, where the low oxygen level protected iron in the rock from rusting.

By 360 million years ago, the Caledonide mountains had been worn to stumps, and Britain had drifted north to sit on the equator. The global sea level had risen, and much of our landmass was covered by a tropical sea. For the next 35 million years, these warm waters created some of our most beautiful and charismatic landscapes. One of them is especially close to my heart.

A Taste of the Tropics

I was born and grew up in Ilkley, on the edge of the Yorkshire Dales National Park. This was the playground of my youth, where I climbed the three peaks of Ingleborough, Pen-y-Ghent and Great Whernside, and marvelled at the untouched beauty of places such as Malham Cove. Sunday walks offered a choice of 'woods, moors or river', and when Dad got our first car – well, a van, actually – the Dales were opened up to us.

I always loved the ancient order of the hilly fields and low-lying meadows, framed by all those dry-stone walls. The walls, which are painstakingly built by hand from the broken fragments of rock that litter the ground, are at the heart of this landscape. And it's from these walls that you can fathom exactly what made the landscape so special. Turn over a few of the rocks and you might find something surprising: fossilized coral. Coral in the heart of Yorkshire? It seems unbelievable when you're hundreds of feet above sea

Yorkshire was one tiny slice of a vast tropical sea.

(ABOVE) THE SOFT LIMESTONE LANDSCAPE OF THE YORKSHIRE DALES WAS CREATED IN EQUATORIAL CORAL SEAS.

level, trudging against a freezing northerly and dreaming of supping a pint by the fireside of an Ingleton pub. But 360 million years ago, this part of Yorkshire was one tiny slice of a vast tropical sea. The newly formed Britain was close to the equator, baked by a tropical sun and covered with seas that teemed with life. Corals flourished in the clear, shallow water and built huge reefs of limestone, just as they do in the tropics today. It is these ancient reefs that have been etched into the soft limestone landscape of the Dales.

Constructed in water, the Dales have also been shaped by water. The rain here falls in prodigious quantities, giving birth to a network of great rivers, such as the Wharfe, the Nidd, the Swale and the Ure. Rain, being slightly acidic, eats away at limestone, dissolving the alkaline minerals within it. The countless streams that run off the Dales slowly carve their way through the rock, creating gorges and waterfalls whose names read like a who's who of beauty spots: Aysgarth Falls, Malham Cove, Gordale Scar, Thornton Force. The streams don't just cut through the rock – they disappear into it. The ground is riddled with potholes and caverns carved out of the limestone by underground streams and rivers.

Hidden under the rocky slope of Ingleborough, Gaping Ghyll is Britain's largest underground chamber. At its deepest point is a pool fed by Britain's tallest unbroken waterfall, a cascade 364 ft (111 m) high. The water disappears into sinks in the cave floor and flows through more than 2 miles (3 km) of subterranean passageways before emerging near the village of Ingleton. And this is just one of hundreds of underground streams and rivers that percolate

(RIGHT) GAPING GHYLL, BRITAIN'S LARGEST CAVERN, IS SAID TO BE BIG ENOUGH TO ACCOMMODATE YORK MINSTER.

through the hills and are rumoured to pass right under the highest peaks.

For 35 million years, tropical seas covered a large stretch of Britain. The Yorkshire Dales are not the only part of the landscape that formed in those warm waters 360 million years ago. The Derbyshire Dales, parts of the Lake District and southern Scotland, and the Mendips – with their most famous attraction, Cheddar Gorge – all formed from the same type of rock: Carboniferous limestone. But the largest mass of Carboniferous limestone is the one that stretches across Ireland. On Ireland's west coast, the Burren National Park looks very similar to the Yorkshire Dales, with dry-stone walls, potholes and 'limestone pavements' – outcrops of bare rock that have weathered into rounded slabs and deep crevices. As in the Dales, the Irish limestone pavements are inhabited by some of our rarest flowers, such as the spring gentian and mountain avens.

I always remember the journey back to Ilkley after a day out in the Yorkshire Dales. I'd stare out of the car window, my eyes following the jigsaw of rocks in the dry-stone walls that hugged the road home. As we approached Ilkley, the rocks slowly changed from the light limestone to a much darker, grittier rock. These subtle changes in the walls paralleled a change in the rock underfoot. Over the course of a couple of miles, we had left the limestone behind and entered a new and much bleaker landscape – the gritstone moors of the Pennines, including Ilkley Moor, which overlooked my home town.

Around 325 million years ago, the coral reefs of the early Carboniferous Period began to disappear. A new chain of mountains had risen to the north, and the cycle of erosion and deposition had begun again, just as in the Devonian Period. As the mountains eroded, rain and rivers flushed out the rocky debris and dumped it at sea, smothering the reefs with sediment and slowly turning the shallow sea into river deltas and land. Later, the sediment was compacted into a hard, unforgiving sedimentary rock that now forms the backbone of England: the Pennines.

The rock is called gritstone, and it's easy to see why. On close inspection, you can see that it's made of lots of small pieces of sand, grit and pebbles cemented together. In fact, it looks just like the sand and gravel that you find on the bank of a large river, and that's exactly where much of it formed. Many of the gritstone outcrops and escarpments dotted across the Pennines – such as Stanage Edge near Sheffield, and the 'Cow and Calf' rocks that overlook Ilkley – are the remains of ancient river banks, though the rivers that once flowed here would have dwarfed anything in the Pennines today. Some 300 million years ago, this area would have looked more like the Mississippi or Niger deltas, with vast, meandering river channels dumping their sediment as they flowed slowly into the sea.

The moorlands on the flatter tops of the Pennines are a world away from the botanically rich limestone meadows of the Dales. Unlike limestone, gritstone is impervious to water, so rain just sits on the surface, creating a boggy, acidic soil that is low in nutrients. Only hardy, acid-tolerant plants, such as heather, can survive here, and they tend to dominate. But 300 million years ago, things were very different. As the shallow seas filled with sediment and turned into swampy deltas, vegetation took hold in a big way. The late Carboniferous (*see box* 'Geological Periods', p. 28) saw the rise of the greatest forests our islands have ever known – forests that left a rich legacy that would power a nation: coal.

The great seams of coal in South Wales, the Midlands, Yorkshire, Northumberland and southern Scotland are the fossilized remains of ancient tropical forests. If you're lucky, you can sometimes see the

(OPPOSITE) IN VICTORIA PARK, GLASGOW, THE FOSSILIZED STUMPS OF A FEW CARBONIFEROUS TREES STILL STAND.

Geological Periods

Millions of Years Ago	Geological Period	Key Events Formed	Landscapes
4600–570	Precambrian	The continents and oceans form. Life is very basic (bacteria) until late in the Precambrian.	Outer Hebrides, Anglesey
570–510	Cambrian	Scotland and England are on separate continents, 3000 miles (5000 km) apart. The seas are rich in life, especially trilobites.	Snowdonia, Lake District, Scottish Highlands
510–439	Ordovician	Continental drift brings England and Scotland together. Welsh slates form. Britain is covered by deep seas.	Wales
439–408	Silurian	Scotland and England collide. Scotland's great mountain range (the Caledonides) forms.	Wales, Shropshire
408–362	Devonian	Fish are the dominant form of life. Old Red Sandstone continents form as the Caledonide mountains erode.	Orkneys, Herefordshire, Black Mountains
362–290	Carboniferous	Britain drifts north of the equator. In the Early Carboniferous, a tropical sea covers Britain and limestone landscapes form.	Yorkshire Dales, Burren, Mendips
		In the Late Carboniferous, the sea subsides, and lush, swampy forests (which will later become coal) cover the land.	Pennines, South Wales, Northumberland, southern Scotland
290–245	Permian	The continents join to form a single 'supercontinent' called Pangaea. Britain, now a desert, is at the same latitude as the Sahara.	West Midlands, southern Scotland, Devon
245–208	Triassic	Desert conditions continue. The land is flattened by a period of great erosion. The first dinosaurs emerge.	Midlands, Trent Valley, south Devon coast
208–145	Jurassic	A tropical sea covers Britain again in the Early Jurassic. Beds of limestone and the fossils of huge marine reptiles are laid down. In the Late Jurassic, the sea recedes and dinosaurs become common.	Dorset, Cotswolds, North Yorkshire
145–65	Cretaceous	Dinosaurs continue to flourish, as do rich tropical forests of cycads and monkey puzzles.	Dorset coast, Kent, Sussex, Isle of Wight
		In the Late Cretaceous, Britain is submerged yet again. Chalk deposits form from microscopic plankton. The dinosaurs disappear.	South and North Downs, Chilterns, Salisbury Plain
65–1.64	Tertiary	In the Early Tertiary, the Atlantic Ocean is born. Northern Ireland and western Scotland are the scene of much volcanic activity.	Giant's Causeway, Skye, Mull
		Subtropical swamps flourish in the Late Tertiary. Mammals become the dominant animals on land.	London, Hampshire

impression of a tree trunk or a leaf in the coal itself. The Carboniferous was the first time in the Earth's history that plants dominated the land, and dominate they did. The trees were giants, growing to 160 ft (50 m) tall, but they were very different from the trees of today's tropical forests. Their closest living relatives – horsetails and the tiny club mosses, which are just a few inches tall – are now small understorey plants. But in Carboniferous times they grew to giant proportions and made up the forest canopy.

Coal forms from peat, and peat can form only in stagnant water, where the lack of oxygen prevents dead vegetation from rotting. The Carboniferous forests were lush and very swampy, providing ideal conditions. Dead vegetation sank in the stagnant water and built up to form peat. When the sea level rose, as it did periodically in the Carboniferous era, mud and sand were dumped on top of the forest, permanently trapping the peat. When the sea subsided again, forests grew back and the cycle repeated. Over millions of years, the trapped peat became compacted into the coal that we mine today. It takes about 30 ft (10 m) of peat to make a seam of coal just 3 ft (1 m) deep, so you can imagine just how rich the Carboniferous forests must have been.

Occasionally, a few of the great trees were buried by a mudslide and became fossilized as mudstone rather than coal, preserving their shape. At Victoria Park in Glasgow, a fragment of the ancient Carboniferous forest survives in the form of mudstone tree stumps – a rare glimpse into our past.

There are other fossils besides plants in Britain's Carboniferous deposits. In the bands of mudstone sandwiched between coal seams are some familiar-looking animals, though on a scale normally seen only in horror movies. Like the trees, the dragonflies, millipedes, cockroaches and scorpions of the Carboniferous were absolutely enormous. The scorpions were 2 ft (60 cm) long, dragonflies had wingspans of nearly 3 ft (1 m), and one species of millipede would have been big enough for a human to sit on. At the top of the food chain were 'temnospondyls' – predatory amphibians that looked like 7-ft (2-m) long salamanders.

By 290 million years ago, the coal swamps and forests on which our nation was built were starting to wane. As the British Isles drifted north and away from the equator, things started to become a little parched.

Desert Britain

Around 290 million years ago, Britain suffered another continental collision. A colossal continent called Gondwana, which had sat over the South Pole for millions of years, was now moving north – and the continent called Laurasia, which contained our islands, was in its way. Between the two was the shrinking Rheic Ocean, which had smothered much of Britain during the early Carboniferous Period, creating the landscapes of the Dales.

When Gondwana struck Laurasia, the two formed a gigantic 'supercontinent' called Pangaea, incorporating nearly all the land on the planet. Devon and Cornwall bear the scars of this continental collision in their craggy coastlines. In the cliffs at Hartland Quay in north Devon, what were once horizontal beds of rock have been bent and buckled between the jaws of a continental vice. And on the south coast of Cornwall, Lizard Peninsula is an ancient block of oceanic crust that was thrust skywards as it was compressed between the continents.

The wild moors of Devon and Cornwall formed during the collision. As the continents pushed together, they squeezed the crust into a dome of rock, cracking it in the process. Into these cracks crept magma, forced by pressure beneath. This molten rock didn't erupt through volcanoes; instead it was confined in domes of magma a few miles under the surface, where it slowly cooled and became granite. The softer surface rock has been steadily eroded since then, leaving the tips of the granite plugs exposed as Dartmoor, Bodmin Moor, Land's End and the Scilly Isles.

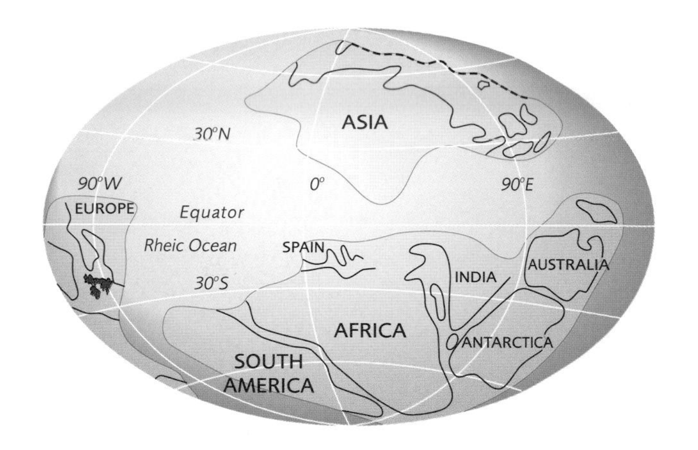

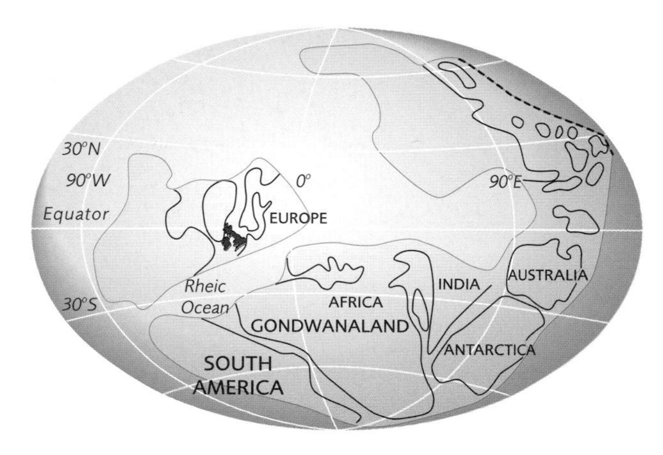

(ABOVE) SOME 290 MILLION YEARS AGO, THE CONTINENTS JOINED TO FORM A SINGLE LANDMASS, CALLED PANGAEA, WITH BRITAIN TRAPPED IN THE CENTRE.

When Pangaea formed, Britain became marooned in the centre of an immense landmass, far from the humid ocean air and at a similar latitude to the Sahara. For much of the next 80 million years – the Permian and Triassic Periods – Britain was desert. Crescent-shaped sand dunes smothered the land and marched east to west, blown by the wind, just as in the Sahara today. It's strange to think of Birmingham as a desert, but just to its west is the proof: massive fossilized sand dunes. You can still see their slopes in the rich, crumbly sandstone.

In a quarry near the village of Hopeman, on the southern shore of Scotland's Moray Firth, similar sandstone is excavated for building material. Walk around any of the local towns and villages, and you can't help but notice the beautiful yellow sandstone from which the buildings are made. Over the years, quarry workers had occasionally unearthed what look like reptilian footprints in this rock, but they had

(RIGHT) THE BENT AND BUCKLED CLIFFS OF HARTLAND QUAY ON THE NORTH DEVON COAST SHOW THE POWER OF THE CONTINENTAL COLLISION THAT TRAPPED BRITAIN AT THE HEART OF PANGAEA.

never found actual fossils. Then, in 1997, a quarry worker noticed a hole about the size of a 10 pence piece and alerted a local palaeontologist. The hole looked like an air pocket, but air pockets in solid sandstone are highly unusual. So they cut out the block of rock around the hole and took it to a hospital to examine its interior with a brain scanner. As the block was scanned layer by layer, an incredible image was unveiled on the monitor. The cavity was shaped exactly like the skull of a prehistoric reptile called a dicynodont – a tusked herbivore, about the size of a sheep, that lived before the age of dinosaurs. This animal was a desert specialist and probably moved in herds across the Permian dunes of northern Scotland. Dunes and dicynodonts aren't the only evidence that Britain was once part of a great desert. Stretching under Cheshire are huge reserves of salt which brought riches to the Romans and to modern petrochemical giants, such as ICI. In the late Permian, 245

It's strange to think of Birmingham as a desert, but just to its west is the proof: massive fossilized sand dunes.

(ABOVE) THE HUGE DUNES THAT SMOTHERED BRITAIN DURING THE PERMIAN ARE EVIDENT IN THE SLOPES AND LINES ON THE WALL OF MAUCHLINE SANDSTONE QUARRY IN SOUTHERN SCOTLAND.

million years ago, the limbs of shallow seas reached into Pangaea's arid interior and flooded this part of Britain. The water evaporated under the relentless heat of the sun, leaving behind parched salt flats – a landscape more like the salt pans of Bolivia than modern Cheshire. The underground salt has now been mined in such quantities and for so long (more than 2000 years) that the land has subsided, leaving a lowland of scapes and meres that have become havens for wetland wildlife.

Jurassic Seas

The giant continent of Pangaea started to break up in the early Jurassic Period, causing the sea level to rise and Britain to disappear under water again. The Jurassic seas were awash with a whole new wave of animals, and their remains can be found among the sedimentary rocks of Dorset's 'Jurassic Coast', England's only natural World Heritage Site. Over the centuries, an army of fossil hunters has combed the beaches and cliffs of Dorset, unearthing an astonishing variety of ancient creatures. Crack open any likely looking rock that has been prized from the cliff by a storm and you have every chance of uncovering a fossil. There are ammonites and belemnites, ichthyosaurs and

(ABOVE LEFT) WHEN SCIENTISTS PUT A PECULIAR BLOCK OF SCOTTISH SANDSTONE UNDER A BRAIN SCANNER, THEY DISCOVERED A CAVITY SHAPED EXACTLY LIKE THE SKULL OF A DICYNODONT – A 250-MILLION-YEAR-OLD REPTILE.

(LEFT) CHESHIRE WAS A REGION OF DESERT AND SALT PANS DURING THE PERMIAN, MUCH LIKE BOLIVIA TODAY.

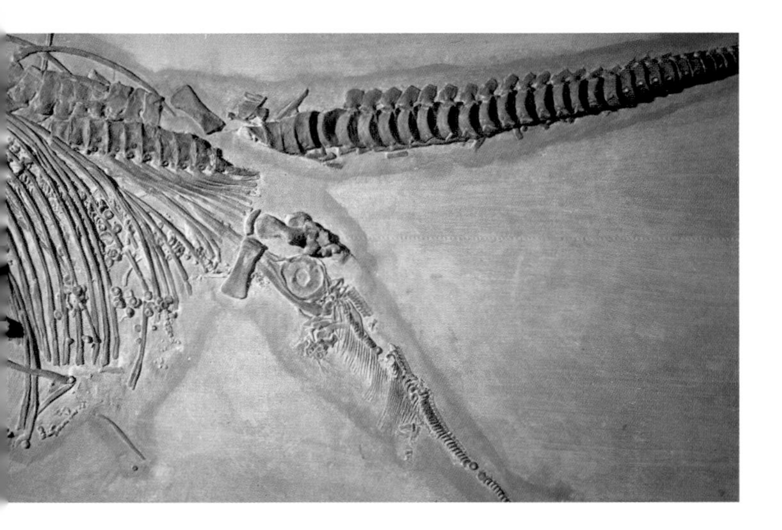

plesiosaurs, all names that have become familiar since the TV series *Walking with Dinosaurs*.

Fossil hunters have even unearthed fossilized vomit, full of the tiny bones of semi-digested fish. It probably came from an ichthyosaur, a marine reptile shaped like a dolphin. Immaculate ichthyosaur skeletons have been found with unborn babies still in their wombs. These animals obviously weren't egg-layers like crocodiles – they gave birth to live young, making them the reptilian equivalent of dolphins.

By the Jurassic Period, Britain's global journey had brought it to about the same latitude as today's southern Mediterranean. The Dorset coast would have resembled the Bahamas, with shallow lagoons and sandbanks surrounded by a warm sea. Once again, the evidence is in the rock. Just as in the Carboniferous era, beds of limestone were forming. As the balmy lagoons basked under the subtropical sun, layer after layer of calcium carbonate was deposited on the sand and reefs – a bit like the inside of a kettle getting caked in a little more limescale with each boil. It was a slow process, but over 60 million years the calcium built up into thick beds of Jurassic limestone. And, as with the Yorkshire Dales, the limestone formed some of our most scenic landscapes. The vales and scarps of Jurassic limestone stretch from Dorset, along the Cotswold ridge, through the Midlands and up to the Lincolnshire Wolds.

Jurassic limestone has long been a popular building material. It's easy to work, very durable and it has an ageless quality, giving endless character to the villages and towns that sit along seams of this rock, none more so than the chocolate-box beauties of the Cotswolds. The purest limestone with the fewest faults is the most highly prized, and no limestone is purer than that from the island of Portland in Dorset. Quarries old and new pockmark this island, which produces the world-famous Portland stone that was used to construct some of our most famous buildings, including St Paul's Cathedral.

Once again, the quarrymen who work Jurassic limestone have found more than just rock in their quarries. In the upper layers, Jurassic limestone is marked by ripples just like those found on a sandy beach after the tide flows out. In other places are clusters of doughnut-shaped swirls – probably the faint remains of fossilized trees. Beaches and forests tell us there was land here some 145 million years ago, at the end of the Jurassic Period. And footprints in the rock tell us that the land was inhabited by dinosaurs. This was our very own Jurassic Park.

Between 160 and 100 million years ago, Britain was rich in dinosaur species. After the Jurassic sea withdrew, swampy forests of monkey puzzles and palm-like cycads flourished on the land. Fossils from all over the south of England (especially the Isle of Wight and the Wealden basin between the North and South Downs) reveal that an impressive range of herbivorous dinosaurs lived among these forests, including the small but fast-moving *Hypsilophodon, Iguanodon* (the one with the hitchhiker's upturned thumbs), and the gargantuan, long-necked *Brachiosaurus*. Some of these may have fallen prey to the

(ABOVE LEFT) SOME ICHTHYOSAUR FOSSILS HAVE BEEN FOUND STILL WITH THEIR UNBORN BABIES.

massive predator *Megalosaurus*, which rivalled *Tyrannosaurus* in stature. Tracks thought to have been made by this giant were recently discovered under a sandy beach on the Isle of Skye.

The dinosaur fossils and footprints disappeared altogether from Britain's geological record around 100 million years ago, some 35 million years before the rest of the dinosaurs were pushed over the evolutionary edge, apparently by a catastrophic meteor impact off the coast of Mexico. Our dinosaurs vanished early because they had nowhere left to go – 100 million years ago, our islands were swallowed by an ocean yet again.

A Submarine World

If you ask any tourist to name one famous natural landmark in Britain, I'll bet that the White Cliffs of Dover would figure most frequently. These imposing chalk cliffs stand like sentinels on the south coast, facing the European mainland. They are famous worldwide, partly for their dramatic location and partly because they look as though they get a regular lick of white paint. Like the Seven Sisters at Beachy Head down the coast, the White Cliffs of Dover hold clues to Britain's fate after the dinosaurs vanished.

Walk along the beach at Dover and you'll find lots of chunks of chalk that have broken off the fast-eroding cliffs. No matter how closely you look at the chalk, you won't see any fossils – chalk is a remarkably uniform rock – just the odd flint. But if you happen to have an electron microscope to hand, you will see something interesting. This rock is in fact made entirely of fossils – billions and billions of them, so small that you could fit thousands on a pinhead. The fossils are the chalk shells of microscopic plankton called coccolithophores, organisms that still exist today, drifting aimlessly in the sunlit surface of the sea, where they form the base of the marine food chain. When their short lives end, coccolithophores fall like snow to the ocean floor, their shells building up in layers of chalky ooze. This is exactly what happened in Britain during the Late Cretaceous. The planktonic 'snow' fell for more than 35 million years, slowly accumulating and compacting into huge beds of chalk. There are no land-derived sediments in this chalk, so Britain must have been utterly swamped by a rise in sea level. Only our highest peaks might have had their heads above water.

Britain's chalk landscapes stretch inland from the Sussex and Kent coasts, forming the North and South Downs, the Chilterns, the plains of Salisbury and Wiltshire, and up through Norfolk to the huge cliffs of

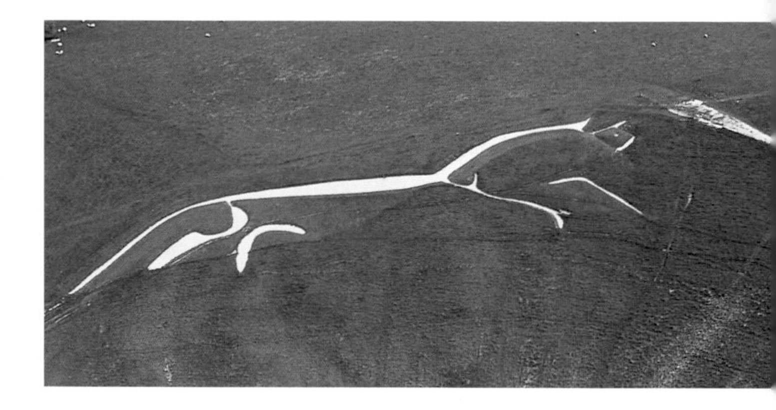

Flamborough Head in Yorkshire. This landscape is riddled with trout-filled streams, the chalk-filtered water running crystal clear. It is also a landscape of grasslands, grazed for millennia by sheep, which keep the grass short and encourage a rich spectrum of wild flowers, including many orchids. The flowers, in turn, provide food for some of our rarest butterflies.

Humans have a very long association with Britain's chalk lands. Embedded in the chalk are nodules of flint, a much harder rock that splinters into sharp fragments.

(ABOVE) THE WHITE HORSE AT UFFINGTON IN OXFORDSHIRE. OUR CHALK LANDS HAVE A LONG HISTORY OF SETTLEMENT, REVEALED BY THE MANY ANCIENT MONUMENTS SCATTERED AMONG THEM.

Flint was highly sought after in the Stone Age, when it was used to make arrowheads, axes and many other tools. These areas were also some of the first to be farmed, with the rich and easily worked soils making them so attractive. Scattered among the low hills and plains are some of our oldest monuments, including Stonehenge and the Uffington White Horse.

By 65 million years ago, Britain had drifted north to the latitude of modern Spain. The final leg of the journey was about to start with a bang.

Birth of an Ocean

Legend has it that the Giant's Causeway in Northern Ireland was built by a giant called Finn McCool. According to one of many local yarns, Finn McCool made a bridge of stepping stones all the way to the Hebridean island of Staffa so that he could challenge a Scottish giant to a contest of strength. When Finn saw how much bigger the Scottish giant was, he turned and ran back, smashing the bridge up behind him.

The Causeway's true origins are more spectacular and violent than the myth. The hexagonal stone columns, which descend into the sea like a staircase, are made from basalt – solidified lava. Around 60 million years ago, Northern Ireland was a volcanic hotspot. Lava forced its way up through fissures in the ground and flooded onto the surface. As the lava cooled and hardened, it shrank and split along a lattice of cracks, forming hexagonal fragments. The same process produced the famous wall of basalt pillars around Fingal's Cave on the island of Staffa, and there are similar structures on the northeast coast of Skye.

All these volcanic events were caused by an immense rift in the Earth's crust. The continent that had once joined Europe and North America was now tearing apart, its two halves pushed in opposite directions by lava welling up from below, a process that continues to this day. This was the moment the Atlantic Ocean was born. And Northern Ireland, like present-day Iceland, sat perilously close to the rift.

There is more evidence of this volcanic era among the Hebrides of western Scotland. The magnificently wild mountains of Skye – the Cuillins – formed from lava flows nearly half a mile thick.

(ABOVE) THE WHITE CLIFFS OF DOVER ARE MADE OF BILLIONS AND BILLIONS OF MICROSCOPIC PLANKTON CALLED COCCOLITHOPHORES, WHICH HAVE SLOWLY COMPACTED INTO HUGE BEDS OF CHALK.

The mountains of Mull are also made of lava, since carved into cone-shaped hills and rounded valleys by Ice Age glaciers. And so are the remote islands of St Kilda and Rockall, of shipping forecast fame.

As well as giving birth to the Atlantic, the volcanic activity had another important effect. A huge well of lava built up beneath the crust and, like a boil under the skin, it pushed the crust up, lifting the sea bed clear of the water. Britain had been submerged for 40 million years, covered by the chalk seas of the Late Cretaceous. But now, as the Atlantic was opening up, the islands emerged from the water for the final time. Our journey was nearly complete.

Over the next 40 million years (the Tertiary Period), Britain's youngest rocks formed, mostly in the southeast. London sits on a bed of soft, unconsolidated clay, which hides fossils of a familiar and recent addition to the evolutionary tree: mammals. By the time this clay formed, Britain had reached its current location, on the edge of an ever-widening ocean. It had completed an epic journey across the globe, each leg of which had added another layer of bricks to our foundations, another layer of the rocks that form our land. The rocks were now in place, but the landscape and coastline were far from complete. It took the events of the next 3 million years to finally carve the British Isles into shape.

(RIGHT) THE WALLS OF FINGAL'S CAVE ON THE SCOTTISH ISLAND OF STAFFA ARE MADE FROM HUNDREDS OF BASALT PILLARS, MUCH LIKE THE HEXAGONAL 'STEPPING STONES' OF THE GIANT'S CAUSEWAY IN NORTHERN IRELAND. THESE BASALT STRUCTURES FORMED FROM FLOODS OF LAVA RELEASED AS THE EARTH'S CRUST SPLIT APART.

The Big Freeze

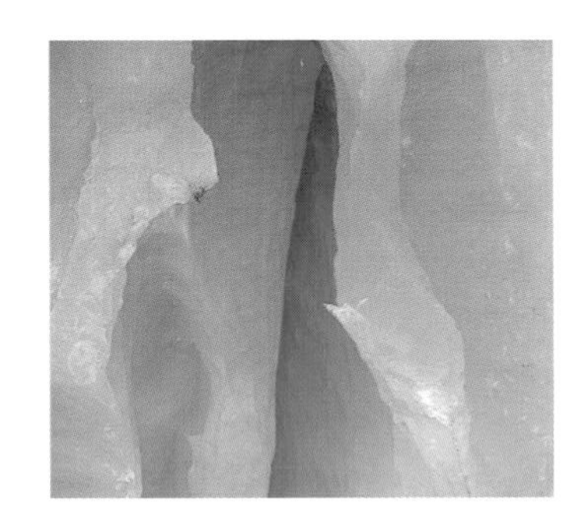

3 million to 14,000 years ago

THREE MILLION YEARS AGO, AFTER A STRING OF CATACLYSMIC EVENTS, BRITAIN WAS SETTLING DOWN. NO LONGER WERE IMPOSING VOLCANOES SPEWING STEAMING LAVA INTO THE AIR. NO LONGER WERE MAJESTIC MOUNTAINS BEING THRUST UPWARDS FROM THE EARTH'S CRUST. AND NO LONGER WERE DINOSAURS STALKING THE LANDSCAPE. IN COMPARISON WITH ITS EARLIER HISTORY, BRITAIN WAS FOR ONCE RELATIVELY CALM – WE HAD REACHED OUR PRESENT GEOGRAPHICAL POSITION, OUR UPLANDS WERE ALL IN

PLACE, AND THE MAMMALS THAT SUPERSEDED THE DINOSAURS WERE ENJOYING LIFE TO THE FULL. ALL IN ALL, THINGS WERE TICKING ALONG NICELY.

BUT THIS TRANQUILLITY WAS ABOUT TO BE DISRUPTED FOREVER. JUST AT THIS TIME A SEEMINGLY INSIGNIFICANT EVENT THOUSANDS OF MILES AWAY ON THE OTHER SIDE OF THE WORLD SET OFF A CHAIN REACTION THAT WAS TO CHANGE PROFOUNDLY THE FACE OF BRITAIN: THE ICE AGE WAS ABOUT TO BEGIN.

The Big Chill

The impact of the Ice Age on Britain is impossible to exaggerate. It was vast. It was monumental. It was colossal. Nothing here today comes even remotely close to conditions felt during the coldest years of the Ice Age. Our most severe winter in recent years was in 1963, when it was so cold that parts of England were covered by snow for 10 weeks solid. At the seaside town of Eastbourne – famed for its warm and sunny weather – the sea froze for about 100 ft (30 m) from the shore. In Oxford, ice on the Thames was so thick that cars were able to drive across, while boats and barges sat marooned, trapped in the river's icy grip. More than 200 of London's usually dependable buses broke down, their diesel frozen firm by the biting cold. And at home in Ilkley, folk in the surrounding villages were cut off by giant snow-drifts, and my dad made us a sledge that lasted for 20 years.

A winter in which sea ice forms on the usually balmy south coast and the Thames freezes over sounds desperately harsh, and certainly for those of us around at the time it was a chilling experience. But even the coldest day of 1963, when temperatures fell to −15°C (5°F), was nothing compared to an Ice Age winter.

Two million years ago, winter temperatures would hardly have registered on our thermometers. And there were bucket-loads of snow – more snow than we have ever seen since. It was so cold that the snow didn't melt, not even in summer. Instead of melting, it became deeper and deeper, and each new layer increased the pressure on the snow underneath. As this pressure steadily rose, the snow turned to ice, and over many years, this layer of ice became thicker and thicker.

At the coldest point of the Ice Age, 500,000 years ago, a colossal ice sheet stretched all the way from the

North Pole to London. There was ice all the way across Wales and most of Ireland, and ice across East Anglia. At times, virtually the whole of Britain was buried under a sheet of ice up to a mile and a half deep. Not even the top of Ben Nevis, Britain's tallest mountain standing 4409 ft (1344 m) high, would have poked through.

The extreme conditions were certainly not restricted to these shores – across the northern hemisphere there were ice sheets in an almost continuous belt. From Britain the ice spread eastwards across Scandinavia to Siberia, China and North America. The whole of Greenland was buried by ice, and the Arctic ice cap was bigger than it has ever been since.

The climate was bitterly cold, with temperatures as low as –80°C (–112°F). Wind would have whipped its way down from the north, gusting at up to 200 mph (320 km/h), and when the wind died down it must

(ABOVE) THE FROZEN SEA AT ST ANNE'S, LANCASHIRE, FEBRUARY 1963.

(OPPOSITE) NORWAY IS ONE OF THE FEW PLACES WHERE YOU CAN REALLY UNDERSTAND HOW MONUMENTALLY MASSIVE THE ICE SHEET IN BRITAIN MUST HAVE BEEN.

(PREVIOUS PAGE) THE IMPACT OF THE ICE AGE ON BRITAIN WAS IMMENSE.

have been eerily silent. Britain was literally an icy desert where nothing could survive, but in spite of the silence up above, in the heart of the ice all hell appeared to be breaking loose.

When you have an ice sheet 1 mile (1.5 km) thick, which is so vast that it covers almost the whole of Britain, it's difficult to believe that it can move. But that is exactly what glaciers and ice sheets do. They travel. So slow is their progress that if you were to spend an entire day sitting next to one, you might not notice anything. If you're lucky, perhaps a great frozen chunk might split off and crash spectacularly to the ground below. Other than that, there would be no visible sign of movement. But there is a way you can tell if ice is moving: you can hear it. It fizzes and pops, squeaks and grinds. Every glacier or ice sheet still around today makes noises. Sometimes they sound like the gentle singing of a courting humpback whale; at other times they are more like an old-fashioned ship in full sail, creaking and groaning under the pressure of the wind and waves.

The reason ice sheets move is that the level of snowfall across them varies. Typically, more snow falls at higher elevations, so this is where the ice grows thickest. At the opposite end, the ice sheet is slowly melting or large chunks are falling off. The ice sheet creeps along to this thinner end, replacing the ice that is melting.

It is this movement of ice that has shaped our landscape more than anything else except ourselves. Mobile sheets of ice act like giant bulldozers, capable of obliterating almost anything in their path. But the ice is not acting alone. As it slips along, it repeatedly melts and refreezes, picking up chunks of rock from the ground. The layer of rocky debris in the base of the ice sheet acts like a gigantic piece of sandpaper, eroding, scratching, gouging and polishing the land as it moves.

Surprisingly, it was scratch marks on rocks that first made people aware that the Ice Age had occurred at all. In the 1830s, some rock scratches in plains beyond the Alps were spotted by a Swiss scientist, Louis Agassiz. (He had also noticed similar marks in Scotland.) On closer inspection, he realized that the only thing capable of causing such deep and regular groves in solid rock was the movement of glaciers. Since he was many miles from the Alps, he proposed that there had once been a 'great ice age' in which the glaciers extended much farther. At first the scientific community ignored and even ridiculed him. But he persisted. In 1840 he published *Etudes sur les Glaciers* (Studies of Glaciers), in which he argued that a vast, prehistoric ice sheet 'resembling those now existing in Greenland' had once covered all of Switzerland and much of lowland Europe. Slowly but surely, Agassiz began to win people over, particularly when showing them the rock scratches he had found all over Switzerland, on the tops of mountains and on valley floors.

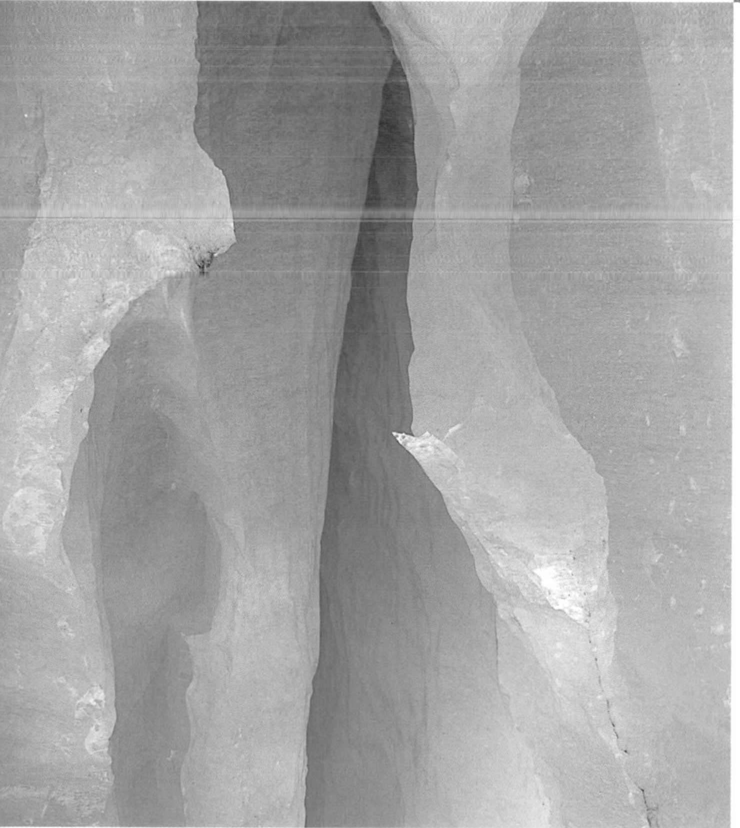

(ABOVE) BEDROCK SCRAPED CLEAN BY ICE – AND NOT JUST ICE. THE GROOVES RUNNING ALONG THE ROCK ARE GOUGES MADE BY ROCKY DEBRIS FROZEN TO THE UNDERSIDE OF A GLACIER, WHICH SLIPPED ALONG LIKE SANDPAPER.

(LEFT) THE HYPNOTIC BLUE ICE IN THE HEART OF A GLACIER.

(OPPOSITE) EVEN OUR TALLEST MOUNTAIN, BEN NEVIS, WOULD HAVE BEEN COVERED FROM BASE TO SUMMIT BY THE IMMENSE ICE SHEET.

Ice does much more than leave scratches in rock. It excavates. It carves out bowl-shaped 'cirques' or 'corries' where glaciers are born. It ploughs its way through V-shaped river valleys, converting them into wide and deep U-shaped ones instead. Often several of these valleys run alongside one another, separated by perilously narrow ridges called arêtes. During the Ice Age, ice scraped its way all across Britain, excavating and sculpting such patterns as it went. The Lake District is one of Britain's most visited National Parks because the scenery is stunning, and it's all here thanks to ice. The famous lakes, with the daffodils that Wordsworth so loved, the towering peaks that inspired Coleridge, and the scree-lined valleys that daring visitors scramble up are all legacies of the Ice Age. There is not a single part of the Lake District that has not been shaped in one way or another by ice, and if, like me, you ever take a rowing boat out on Ullswater, or climb Helvellyn or Great Gable, you've probably marvelled at the raw beauty of the place.

Where there are now small streams, there were once mighty glaciers. Where there are cairns marking summits, there was solid ice extending hundreds of feet higher up. Where there are steep valley sides covered in scree, there were once smooth slopes coated in greenery, until ice had its way. The same goes for Scotland, northern England, Wales and Ireland. Ice is responsible for the shape of much of the British Isles' landscape today. And we put the ice-sculpted landscape to good use. We build rows of electricity pylons and lay railway lines and roads through U-shaped valleys; we dam lakes for hydroelectricity; and above all we enjoy it – we walk, we cycle, we sail, and we climb in a landscape shaped by the Ice Age. Not all our islands' glacial scenery is quite so accessible, though. There are still places too cold or perilous for all but the most adventurous visitors, and they remain a haven for wildlife.

U-shaped valleys and the sharp arêtes between them make spectacular scenery. It's amazing to look at them and think that glaciers once filled them to the brim. Across Britain and Ireland there are also hundreds more subtle remnants of the Ice Age. Strangford Lough in Northern Ireland is one such example. An area of outstanding natural beauty, it's one of the largest sea inlets in the British Isles, with more than 150 miles (240 km) of twisting coastline. Within the lough are 120 small, cigar-shaped islands, but these are no ordinary islands. They are another legacy of the Ice Age: drumlins. Drumlins form when sediment falls from glaciers and gets shaped by the moving ice into long heaps. Their shapes reveal where the ice was moving: the steep end faces the oncoming ice, and the tapering end points in the direction of flow. Other glacial features in the British Isles include moraines (great piles of rocky debris that mark the edge of a glacier, like rubble pushed into a heap by a bulldozer) and tarns (mountain lakes in basins carved by glaciers).

(OPPOSITE) NO PART OF THE LAKE DISTRICT WAS LEFT UNTOUCHED BY ICE. THIS VIEW FROM THE SUMMIT OF GREAT GABLE SHOWS A U-SHAPED VALLEY, WITH WAST WATER IN THE DISTANCE.

(OVERLEAF) BRITAIN WOULD HAVE LOOKED JUST LIKE THIS 500,000 YEARS AGO.

Ice Age Glossary

Arête a steep-sided, sharp-edged ridge formed by two glaciers eroding away on opposite sides of the ridge. Striding Edge in the Lake District is perhaps the most famous example in Britain.

Cirque a semicircular bowl carved out of a mountainside. This is where the snow and ice forming a glacier first accumulate – the birthplace of a glacier. They are common in the Scottish Highlands, where they are known as corries.

Crevasse a large crack in the surface of a glacier.

Drumlin an elongated mound of 'till' (rock debris) dumped and shaped by a glacier.

Erratic a large boulder carried by a glacier and dumped far from its place of origin after the ice melts. Erratics can be found all over the British Isles.

Esker a winding ridge of gravel or sand deposited by a stream that cut a channel through a glacier.

Glacial till jumbled rock debris left by a glacier. Till can include clay, silt, sand, gravel and boulders. Moraines are made of till.

Glacier a large mass of ice that originates on land and slowly flows or spreads.

Hanging valley a valley eroded by a small tributary glacier that joins a much bigger glacier. Because the tributary glacier erodes a shallower trough than the large glacier, its valley is left high up the wall of the main valley after the ice melts. Waterfalls often form at the mouths of hanging valleys.

Horn a pyramid-shaped mountain peak created by several glaciers eroding different sides of the mountain. The Matterhorn is a classic example.

Ice cap a permanent, dome-shaped mass of ice that covers an extensive area (such as the South Pole) and spreads out from the centre.

Nunatak an area of bare land surrounded by ice, such as the peak of a mountain that pokes out of an ice cap. The top of the Matterhorn was once a nunatak.

Ogives bands of darker and lighter ice in the surface of a glacier.

Tarn a small mountain lake, often occupying a basin made by a glacier.

U-shaped valley a valley with steep walls and a broad floor, carved out by a glacier. U-shaped valleys are formed when glaciers remould V-shaped valleys, which have themselves been shaped by rivers.

Jökulhlaup A catastrophic flood that happens when a lake breaches a dam formed by a glacier.

Kame A mound of sand, gravel or other sediment dumped by water released by a melting glacier.

Loess A dusty, wind-blown sediment (or the soil derived from it) that is thought to have formed from rock pulverized by repeated freezing and thawing in the Ice Age.

Moraine a large mound of jumbled debris deposited by a glacier. There are many different types, such as terminal moraines, which are pushed into a heap by a glacier acting like a bulldozer.

(ABOVE) THE MATTERHORN (CENTRE) OGIVE BANDING (RIGHT) STRIDING EDGE IN THE LAKE DISTRICT – A CLASSIC ARÊTE.

Be it shaped by ice or melt-water, erosion or deposition, wherever you live in the British Isles, you will have remnants of the Ice Age somewhere near you. (For names of the weird and wonderful Ice Age features and processes, *see* 'Ice Age Glossary'.)

Beyond the Ice

The ice sheet that covered Britain during the coldest part of the Ice Age was vast, but it did not cover the whole country – parts of southern Britain escaped it. We can tell the extent of the ice by mapping the distribution of sediments that formed beneath it as it melted. Such sediments usually consist of clays containing a scattering of boulders. They tell us that the ice came to an abrupt end about where the M4 now is. In other words, the great ice sheet never stretched further south than Bristol or London.

In London we can be much more precise. Digging up clay for analysis in the heart of a city would be very

The ice sheet that covered Britain during the coldest part of the Ice Age was vast.

(ABOVE) CLEW BAY IN COUNTY MAYO IS A LANDSCAPE OF DROWNED DRUMLINS – STREAMLINED PILES OF SEDIMENT DUMPED BY ANCIENT GLACIERS. DRUMLINS ARE AMONG THE MOST COMMON GLACIAL LANDFORMS IN IRELAND.

difficult, not to mention costly, but thankfully the job was done for us in the late nineteenth century when the London Underground was built. And the London clay revealed a fascinating story: Edgware, High Barnet, Cockfosters, Stanmore, Tottenham and West Hampstead were all covered by ice. But travel south on the Metropolitan Line and you come to the first station that was ice-free: Finchley Road. This unassuming station in north London has a backdrop of Victorian houses and modern supermarkets, but 500,000 years ago it would have had a backdrop far more spectacular. This is where the ice ended.

South of the great ice sheet was a country very different from the England we know today. Although there was no ice, temperatures were bitterly cold – so cold that the soil was frozen solid. Such areas are now called tundra, after the Finnish word for barren or treeless land – and this is exactly what southern England would have been like. With the soil frozen solid for most of the year, most plants don't stand a chance of growing because their roots can't penetrate the ground and they can't extract water. Would-be colonizers also have to contend with an extremely short growing season, biting winds and very low rainfall – less than 5 in (130 mm) a year. Yet, amazingly, there are plants that not only cope but thrive in such conditions. Cotton grass, dwarf heaths, and some species of sedge, moss and lichen are all adapted to withstand sweeping winds and frozen soils. They can carry on photosynthesizing at low temperatures and in low light intensities.

Today tundra covers 15 per cent of the world's land, stretching across the northern hemisphere from Alaska to Canada, from Scandinavia to eastern Russia. Most of the plant species that survive the harsh conditions are the same ones that would have lived in England during much of the Ice Age. But the animals were very different, and evidence of this can be found off the British coast in the North Sea.

One wet blustery day in 1879, a Dutch fishing vessel was bouncing its way across a lumpy North Sea. The fishermen were heading back to Rotterdam at the end of a gruelling week. As they made their way home, they continued to cast their nets, hoping for a final catch. But what they caught surprised them all. Instead of skate or herring, when they heaved the nets aboard, they found that they were full of bones. Massive bones. They clearly didn't belong to fish. Nor did they look like the bones of seals or whales, which the fishermen would have recognized. When they arrived in harbour, the fishermen called the local doctor, who, after much deliberation, pronounced them to be elephant bones. Here at last, he declared, was evidence of Noah's flood. Everyone fell to their knees in prayer.

We now know that the truth is a little different. Although the bones do look like those of elephants, palaeontologists have been able to determine that they are in fact the bones of woolly mammoths. With the development of modern trawlers, more and more bones have been fished up from the floor of the North Sea. They are found so often that returning boats are often greeted by collectors, palaeontologists and entrepreneurs, all eagerly hoping for something special to add to their collection. The remains of an estimated 1 million mammoths have now been trawled from the sea bed, and it's not just mammoths that have been found. There are woolly rhinos – great 3-ton battering rams of fur; giant bison with immense horns; deer with antlers 12 ft (4 m) across; and reindeer and horses like those we see today. And those are just the herbivores. There are also bones of predators, including hyenas, wolves, lions, bears and the fearsome sabre-tooth cat.

Bones of these animals are also found in southern England. When foundations for houses are being dug, or when new cave systems are explored, weird and wonderful bones are often found, helping us to understand what life was like in Ice Age Britain. But why are so many found in the English Channel and the North Sea?

One answer might be that the bones were washed there by rivers and slowly drifted into the muddy depths, where they lay untouched until humans came along. But no river has ever been powerful enough to sweep a million mammoths into the sea. The only possible explanation is that there was once dry land where the bones are found. Much of the North Sea, the Irish Sea and the English Channel was dry. The reason, again, is ice.

The growth of ice sheets disrupted the world's water cycle. Normally, the heat of the sun makes water evaporate from land and sea, filling the air with water vapour. The water vapour rises and cools, causing it to condense and fall back to Earth as rain or snow. Rivers then carry the water back to the sea, completing the cycle. But ice interrupts this cycle. Water still evaporates from the sea and falls as rain or snow, but not all of it returns to the sea. During the Ice Age, a vast amount of water became locked up in ice

When they heaved the nets aboard, they found that they were full of bones. Massive bones. They clearly didn't belong to fish.

(ABOVE) FINDING THE LEG BONE OF A WOOLLY MAMMOTH WHILE FISHING IN THE NORTH SEA BETWEEN THE NETHERLANDS AND DOVER.

sheets on land, causing sea levels to fall by more than 300 ft (90 m). There was no such thing as the British Isles – England was connected to France, and Wales was connected to Ireland.

So trawlers are pulling up bones from the sea bed exactly where the animals died thousands of years ago, perhaps when navigating their way across the tundra plains that linked Britain to Europe. It must have been an astonishing spectacle as herds of mammoths and rhinos wandered through Sussex and Devon, Cornwall and Wiltshire, stalked by bears and lions.

It was the melting of the great ice sheets that caused sea levels to rise once more. The dam that had held back the water cycle was broken, and water flooded into the sea. The rising waters submerged tundra, forests, hills and valleys. Perhaps the most spectacular evidence of the change in sea level comes from limestone caves in Majorca and Mexico. Stalactites and stalagmites take thousands of years to

(ABOVE) BEAUTIFUL TUNDRA-SCAPE IN MODERN ALASKA. MUCH OF SOUTHERN BRITAIN WOULD HAVE LOOKED JUST LIKE THIS DURING PERIODS OF THE ICE AGE.

(OPPOSITE BELOW) DURING MUCH OF THE ICE AGE, BRITAIN WAS JOINED TO MAINLAND EUROPE. HUGE HERDS OF MAMMOTHS AND WOOLLY RHINOS WOULD HAVE MIGRATED ACROSS WHAT IS NOW THE ENGLISH CHANNEL AND THE NORTH SEA.

(OPPOSITE ABOVE) THE STALACTITES AND STALAGMITES IN THIS CAVE IN MEXICO FORMED WHEN THE CAVE WAS ABOVE SEA LEVEL. NOW IT IS DEEP UNDER WATER AND CAN ONLY BE EXPLORED BY SCUBA DIVERS.

(OVERLEAF) THE PANAMA CANAL, COMPLETED IN 1914, NOW SEPARATES NORTH AMERICA FROM SOUTH AMERICA ONCE AGAIN.

form, growing from tiny deposits of calcium carbonate that slowly accumulate as mineral-saturated water trickles through a cave. These magnificent structures can only form above sea level – it is simply impossible for them to form under water. Yet the caves in Majorca and Mexico are 65 ft (20 m) below the surface of the sea.

How Did It All Begin?

The Ice Age carved out our landscape. It gouged its way across mountains, diverted many of our rivers, and left virtually no part of Britain untouched. And all of this seems to have been catalysed by a chance event that took place on the opposite side of the world. Three and a half million years ago, continental drift finally brought North and South America together. It wasn't the sort of violent collision that resulted in the uplift of huge mountains, as when India ploughed into Eurasia, creating the Himalayas. This collision was much more gentle – and yet its effects were just as profound.

When the two landmasses collided, the Isthmus of Panama formed, and instead of two continents, there was one. More importantly, instead of one giant ocean, there were now two: the Atlantic and the Pacific. Before the Isthmus of Panama formed, warm tropical water could circulate between these two bodies of water, but now it was rerouted. A warm ocean current was driven north up the east coast of the Americas, and the current carried moist air towards the Arctic.

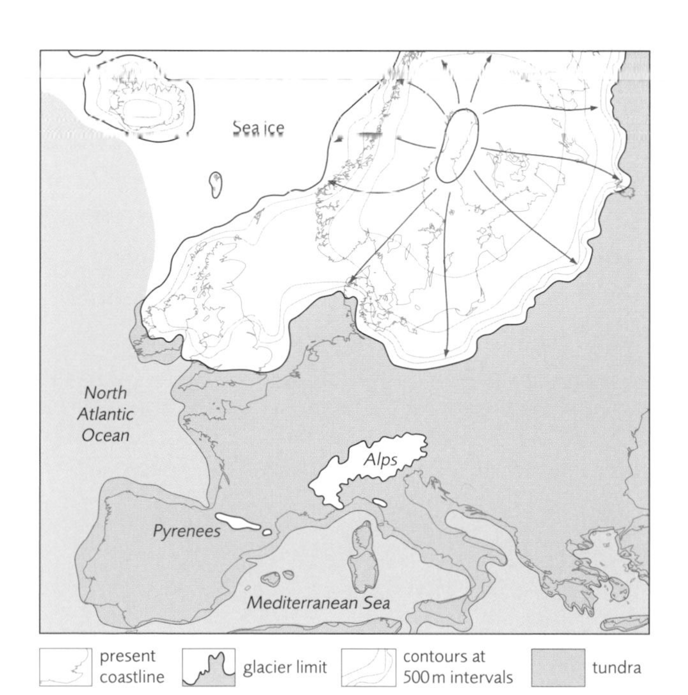

For several million years the Arctic had been not just cold but very dry, with little snow. This was about to change. With the arrival of moist air, snow began to fall in huge quantities. The snow became compacted into ice, and an ice sheet began to grow, forming a vast polar ice cap. Eventually the ice became self-sustaining, acting as a gigantic white reflector that bounced the sun's rays back into space, depriving the Arctic of warmth and so making the region even colder. The Ice Age had started – North America, Europe, Siberia and Asia all began to freeze up.

Hot and Cold

The floor of the North Sea is not the only place where you can find remains of Ice Age animals. One of the best places for bone finds in southern Britain is Trafalgar Square.

When Trafalgar Square was constructed in the 1830s, the builders dug up an impressive collection of bones. Like those found in the North Sea, they were clearly from hefty beasts – there were huge femurs (thigh bones) almost as big as those of mammoths. But no one really knew what the animals were. Years later, when the Ugandan Embassy was being built, more bones were found – and this time people knew exactly what they were. It's rather coincidental that they were found under the Ugandan Embassy, since the bones turned out to be from animals we associate with Africa – hippos, crocodiles and lions.

It seems strange that these animals should have been around in Britain during the Ice Age. Certainly lions could survive the cold – they became larger animals, their greater bulk and longer fur helping them stay warm, and they preyed on reindeer rather than antelope. But hippos? As everyone knows, hippos love the warmth. They spend their days lounging in rivers and muddy wallows, only to emerge at night to feed. Surely their wallows would have frozen over in the Ice Age? Crocodiles, too, need warmth and water, and being cold-blooded they depend on the warming power of the sun to keep their body temperature up. Such animals would not have stood a chance on the tundra. Yet the bones prove they were here, and they were very similar to the species that thrive in Africa today, not some weird subspecies adapted to the cold.

So what were these tropical animals doing in Ice Age Britain? Well, if you thought the Ice Age was one long, uninterrupted cold spell, you'd be wrong. During the 2-million-year period that we call the Ice Age, Britain was not always covered in ice sheets and tundra. For some of the time it was as warm as today, and sometimes it was even warmer. The term 'Ice Age' is actually a bit misleading. In fact, 'Dramatic Change Age' might be a better, if not quite so catchy, description.

Scientists refer to the whole period from the start of the first cold snap 2.6 million years ago to the end of the most recent one some 14,000 years ago as the Ice Age. Within this window, the cold spells are known as glacial phases, and the warmer times are called interglacials. But the point is that there was not just one icy phase but many, all separated by warm periods. In fact, scientists reckon there could have been as many as 50 separate glacial phases during the last couple of million years.

These fluctuations in climate are not fully understood, but the best theory suggests they are caused by complicated shifts in the Earth's orbit around the sun (*see box* 'Ice Age Cycles'). Sometimes the Earth's orbit gives the poles a little more sunlight, and the polar ice caps shrink; at other times the poles get less sunlight and the ice caps grow. The Earth's orbit has probably been wobbling for millions of years, but it was only after the Isthmus of Panama formed that the wobbles had a major effect.

Fifty or so times, ice carved its way across Britain. Some glacial phases were much more severe than

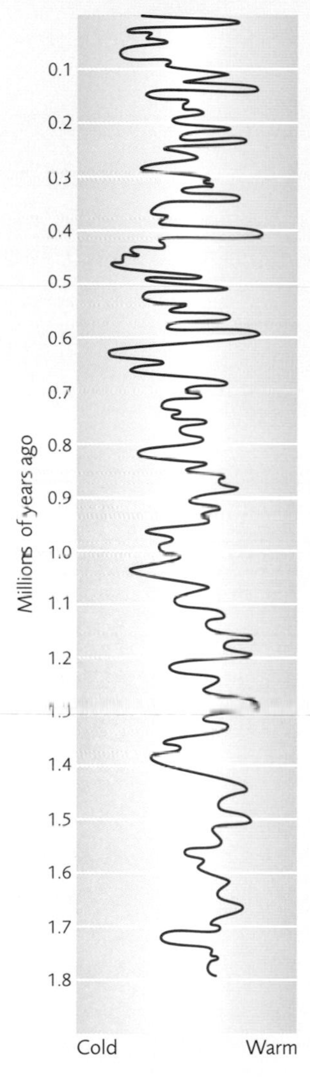

Ever since it was discovered that the Ice Age was not an uninterrupted cold spell but a series of alternating icy and warm phases, scientists have been speculating about what could cause these regular fluctuations. All sorts of theories have been proposed, including changes in ocean currents, volcanic dust in the atmosphere, and meteorite impacts. But none of these satisfactorily explains the regularity of the cycle.

There is one theory that might explain the fluctuations, though. The Serbian astrophysicist Milutin Milankovitch dedicated his career to a mathematical theory of climate, based on changes in the Earth's orbit. According to Milankovitch, subtle changes in orbit cause long-term variations in the amount of sunlight falling at different latitudes. His theory involves three aspects of the Earth's movement:

1 **Eccentricity**. The Earth's orbit around the sun is not circular but slightly elliptical (oval). Over time, it changes from less elliptical to more elliptical, in a cycle that takes about 100,000 years to complete. This cycle coincides almost perfectly with the big glacial phases.

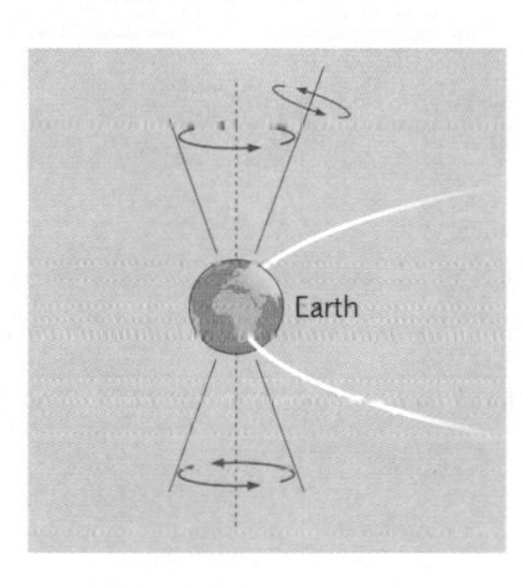

2 **Obliquity**. The Earth spins around like a top as it orbits the sun, and its axis of spin is currently tilted at about 23.5°. This angle varies over time between 22.5° and 24.5°, and it takes 41,000 years to complete one 'obliquity cycle'.

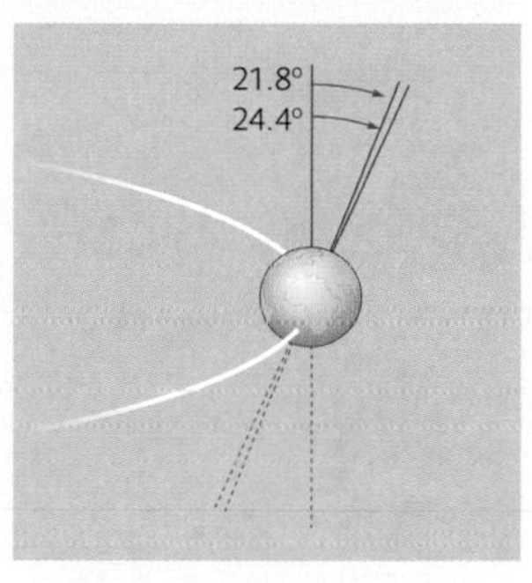

3 **Precession**. The Earth's axis of rotation is itself rotating – just as a leaning spinning top slowly rotates while the main barrel is whizzing round. The Earth's axis goes through a cycle of precession once every 23,000 years.

These three factors combine to cause periodic changes in the level of sunlight falling on the poles, in turn causing the polar ice caps to expand and contract. So far, Milankovitch's theory seems to explain nearly every fluctuation in the Ice Age.

(LEFT) GRAPH SHOWING THE EARTH'S TEMPERATURE FLUCTUATIONS OVER THE LAST TWO MILLION YEARS.
(TOP, ABOVE LEFT, ABOVE RIGHT) FIGURES REPRESENTING THE EARTH'S ECCENTRICITY, OBLIQUITY AND PRECESSION.

others. The biggest, which happened around 500,000 years ago, is known as the Anglian Stage. This was when ice sheets up to a mile and half thick reached as far south as Finchley Road station. Other ice sheets got only as far as North Yorkshire and were a measly half mile thick. But every ice sheet re-carved the landscape afresh, so the stunning arêtes and corries we see in the Lake District, the great U-shaped valleys of Scotland and the drumlins of Strangford Lough are all remnants of the most recent glacial phase, which began some 70,000 years ago and ended only 14,000 years ago.

First Britons

As the ice came and went, so too did the animals. With every shift in climate, there was a mass migration of Ice Age animals. As temperatures plummeted and ice encroached upon the countryside, the warmth-loving animals, such as hippos and crocodiles, as well as elephants and deer, travelled south in search of warmer climes and food, migrating across the continuous land between Britain and Europe. Meanwhile, animals that thrived in the cold, such as mammoths and woolly rhinos, arrived in Britain from the north and east, crossing Scandinavia from areas that were already freezing. Some predators, such as lions and hyenas, stayed whatever the weather and simply switched their prey, but the herbivores had to move on.

One of these migrations brought some of the most successful and feared predators of all to Britain. These dangerous marauders were different from their competitors – they hunted not with sharp claws or huge teeth, but with keen brains. They were humans.

Human occupation of Britain goes back at least half a million years, when *Homo heidelbergensis* lived in southern England. One of the richest sites for evidence of these ancient hominids is Swanscombe, south of the River Thames in Kent, where archaeologists have found part of a female skull and around 100,000 stone 'hand axes', dated to 400,000 years ago. Although Swanscombe has revealed a great deal about the lives of early humans, there is still much that we do not know. Did these people live in caves or build shelters? How many of them were there? How did they catch their food? And why did they disappear? The last question is particularly interesting. It seems likely that a glacial phase drove the early humans to seek pastures new, but when the ice melted and Britain warmed up, few of them returned. And even fewer returned after the next glacial cycle, until there was nobody left. Between 200,000 and 100,000 years ago, Britain seems to have been empty of people.

When people did finally return, they came in the form of a new species: *Homo neanderthalensis*. The Neanderthals are arguably the most famous of our ancestors. Often portrayed as slow, brutish thugs, there were in fact supremely well adapted to Ice Age conditions. They thrived in the cold.

Like the Inuit today, Neanderthals had a short and stocky stature – males were about 5 ft 6 in (1.67 m), and females were 3 inches or so shorter. This squat shape helped them to retain heat, which was vital in Ice Age Britain. Their strange faces may also have helped them survive in the freezing environment. Some anthropologists think their huge, muzzle-like noses, which are often a point of ridicule, may have housed bony projections that warmed and humidified inhaled air, protecting the throat and lungs from being chilled.

Neanderthals were superb hunters. Chemical analysis of their bones has revealed that 90 per cent of their diet was meat, which makes them more carnivorous than wolves. They were strong, quick and powerful. And they were intelligent. Nowhere is this better illustrated than La Cotte de Saint Brelade in Jersey.

A great cliff rises from the beach at La Cotte, and in its base is a rocky shelter that was inhabited during the Ice Age, when the cliff overlooked miles of grassy

Human occupation of Britain goes back at least half a million years.

(ABOVE) LA COTTE DE SAINT BRELADE IN JERSEY, WHERE THOUSANDS OF YEARS AGO MAMMOTHS WERE DRIVEN TO THEIR DEATHS BY THOSE SUPREME HUNTERS – THE NEANDERTHALS.

Neanderthals were intelligent – their hunting methods clearly show that. They also seem to have been capable of speech, judging by the structure of the 'hyoid' bones in their throats. They lived in small groups, and they looked out for each other. In other words, mentally, they had a lot going for them. They were also well adapted physically, with their stocky, powerful bodies. So it seems very strange that they should disappear altogether. A huge debate has raged over exactly what prompted their speedy decline, but it can surely be no coincidence that they perished shortly after we arrived.

Competition

One idea is that Neanderthals simply came second place in the competition to survive. According to Darwin's principle of 'survival of the fittest', it takes only a small advantage to outcompete your rivals. If modern humans were faster, cleverer or more technologically advanced, then Neanderthals would certainly have struggled, just as red squirrels struggled to survive after grey squirrels arrived in Britain. Computer simulations suggest that if Neanderthals had a mortality rate just 2 per cent higher than modern humans, they would have been driven to extinction within only 1000 years.

War

Some people say it was genocide that killed off the Neanderthals. When modern humans arrived, a great war ensued, and we were the victors. But there isn't a great deal of archaeological evidence to support this idea.

Love

Others say it was love rather than war that wiped them out. If modern humans and Neanderthals got on well, they would have interbred. Mixing of the genes could have caused a steady dilution of Neanderthal features until they were lost forever. In 1999, scientists found what appeared to be the arm bone of a Neanderthal–human hybrid in Portugal. However, other scientists have examined ancient fragments of Neanderthal DNA and concluded that interbreeding did not happen – we remained two separate species.

(BELOW) NEANDERTHAL REMAINS HAVE BEEN FOUND ACROSS A SWATHE OF EUROPE AND THE MIDDLE EAST.

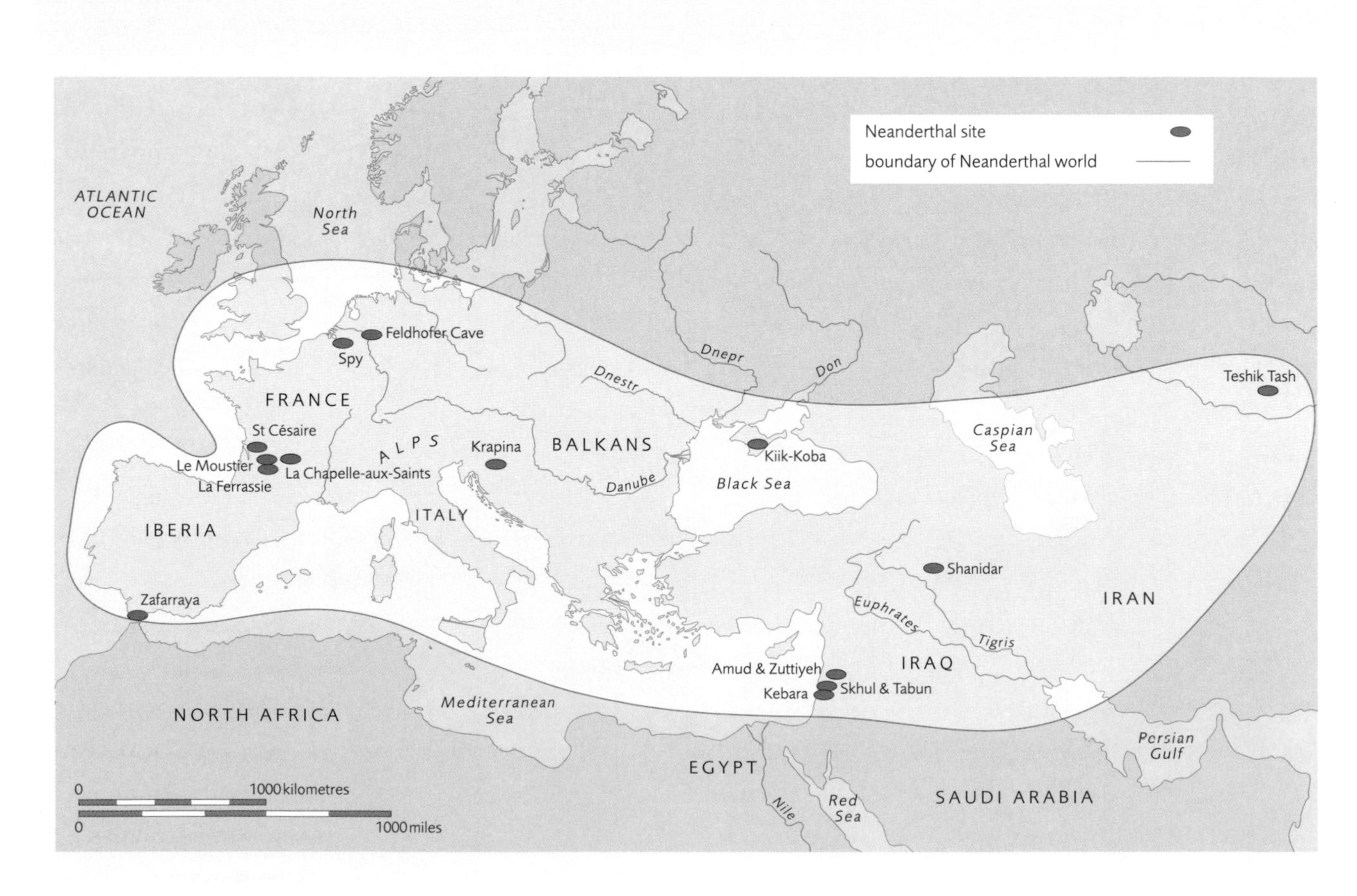

plains rather than the English Channel. When archaeologists dug into the mud at La Cotte, they found Neanderthal stone tools and piles of mammoth and woolly rhino bones, many bearing tell-tale scratches made by their butchers' crude implements. Evidence suggests that the Neanderthals hunted these animals in an ingenious way: by driving them off the cliff. This would have been a dangerous strategy, and one that required planning and teamwork if the animals had to be driven from miles away. When their quarry reached the cliff-top, perhaps the Neanderthals rushed screaming at them, brandishing flaming torches in the dark. In the panic, several of the animals would have lost their footing and tumbled to their deaths. Native Americans used the same technique to hunt buffalo as recently as 100 years ago – it is a complicated, daring and dangerous tactic, and it suggests that the Neanderthals were not simpletons but highly intelligent hunters.

Even so, their days were numbered. Around 30,000 years ago, the Neanderthals died out, apparently replaced by another human species: us. Controversy rages as to why they disappeared – after all, they had co-existed with modern humans for some 20,000 years before the end. There are plenty of theories to explain their demise, but none has been universally accepted (*see box* 'What Happened to the Neanderthals?').

Modern humans made no impact on the landscape for thousands of years, as they struggled to survive in islands dominated by ice. It must have been unbearably difficult. The ice sheet continued to grow after their arrival, reaching its maximum extent

(ABOVE) A RAISED BEACH ON THE ISLAND OF JURA, OFF WESTERN SCOTLAND. AT ONE TIME THIS BEACH WOULD HAVE BEEN AT SEA LEVEL.

around 20,000 years ago. This must have been too much for even the most frost-hardened of humans to cope with, as they seem to have upped sticks and headed south for a while. They didn't return until the climate improved more than 10,000 years later.

Around 14,000 years ago, temperatures began to rise. Perhaps because of some peculiarity in the Earth's orbit, the change was remarkably quick: in less than a human lifetime, the temperature rose from arctic to about today's level. Of course, it still took many centuries for an ice sheet a mile thick to melt, but melt it did. From gentle, drippy beginnings, it was not long before great torrents of water were cascading their way across the country.

The final thaw revealed for the first time the landscape that we know and love today – the majestic peaks, the deep valleys, the sheer arêtes and the stunning lakes. For those humans who ventured back, it must have been an amazing sight.

The End?

The ice left the British Isles long ago, but our islands still bear its scars. If it wasn't for the Ice Age, much of our countryside would not be nearly so beautiful. The Lake District, Snowdonia, the Cairngorms, the Yorkshire Dales – some of our most famous landscapes and National Parks have been sculpted by ice. We canoe in lochs carved by ice, we climb ridges shaped by ice, we drive through great valleys gouged by ice.

But the story is far from over. Even today we are still feeling the effects of the Ice Age. The ice sheet that covered Britain in the last glacial phase weighed billions and billions of tons. So heavy was it that it pushed Britain into the fluid molten layer that all of the Earth's crust sits on. It's hard to believe that solid rock can be squashed like this but that's exactly what happened. When the ice melted and the weight was lifted, the land began to rebound. This explains why

there are 'raised beaches' around Britain's coastline. These beaches were once washed by waves, but now they are up to 130 ft (40 m) above the sea. Parts of Scotland are still rising today, if only by 1 millimetre per year.

And there's no reason to believe that the Ice Age is over. Strictly speaking, we are merely in an 'interglacial' between two glacial phases. The subtle wobbles in the Earth's orbit are still happening, and sooner or later the climate will flip. Temperatures will plunge, snow will start to cover our landscape, and it won't melt in the summer. Slowly the layers of snow will build up to form to a mile-thick sheet of ice that lasts for thousands of years. The scenery we cherish will be re-carved once more, not just in the British Isles but across the northern hemisphere. Sea levels will fall throughout the world, changing coastlines beyond recognition. And the British Isles will no longer be isles.

If it wasn't for the Ice Age, much of our countryside would not be nearly so beautiful.

(ABOVE) THE STUNNING ICE AGE LANDSCAPE OF WESTER ROSS IN NORTHWEST SCOTLAND. ONE DAY THIS BEAUTIFUL SCENERY MAY BE RECARVED BY ICE ONCE AGAIN.

After The Ice

3

14,000 to 8000 years ago

EVERY DAY WELL OVER A HUNDRED TRAINS SPEED THEIR WAY THROUGH THE CHANNEL TUNNEL, MOVING PEOPLE AND GOODS BACK AND FORTH BETWEEN BRITAIN AND CONTINENTAL EUROPE. WE CAN HOP ABOARD AT WATERLOO, AND IN THE TIME IT TAKES TO SETTLE DOWN, READ A PAPER AND HAVE A LIGHT SNACK, WE REACH THE GARE DU NORD IN PARIS WITHOUT EVER GETTING OUR FEET WET. FROM THE ENTRANCE JUST OUTSIDE FOLKESTONE TO ITS EXIT NEAR CALAIS, THE TUNNEL IS MORE THAN 32 MILES (51 KM) LONG,

TWO-THIRDS OF THAT LENGTH RUNNING UNDER THE CHANNEL AT UP TO 130 FT (40 M) BENEATH THE SEA FLOOR. ABOVE THE TUNNEL SIT 100 FT (30 M) OF COLD, GREY WATER. THE TUNNEL WAS OPENED IN 1994, BUT ALTHOUGH WE CAN RIGHTLY CONGRATULATE OURSELVES ON A FINE ENGINEERING ACHIEVEMENT, THIS WAS MOST CERTAINLY NOT THE FIRST TIME BRITAIN HAD BEEN CONNECTED TO EUROPE.

When surveys for the tunnel's construction were being made, evidence of another land link between Britain and Europe was brought into sharp focus. The surveyors discovered a mysterious lost landscape hidden beneath the waters of the English Channel – a sea floor littered with shapes and patterns that could only have formed on dry land. There were river valleys cutting through chalky rocks, sheer cliffs over which powerful waterfalls must have once tumbled, and hills shaped by the action of ancient streams. Today all this scenery is submerged under fathoms of water.

This lost landscape, Britain's very own Atlantis, disappeared under the waves just a few thousand years ago. Back then, from a vantage point high on the White Cliffs of Dover, we would have gazed out not over the choppy waters of the Channel, but over a wide, grassy plain that stretched uninterrupted into France.

The reason for this dramatic difference in view is that 15,000 years ago, Britain was still held firmly in the grip of the Ice Age. The northern half of the country, and much of Europe, was hidden beneath vast, mile-thick ice sheets. Worldwide, there was more than 24 million cubic miles (100 million cubic km) of ice, locking up so much water that sea levels were more than 330 ft (100 m) lower than today – low enough to drain the Channel.

The Big Thaw

But this scene was about to be transformed. Between 14,000 and 13,000 years ago, the big freeze came to an abrupt and dramatic end, and some surprising evidence found in the Lake District helps to unravel the story. The evidence came in the unlikely form of fossilized beetles. Today the species in question are no longer found in the Lakes, but they do still thrive in other parts of the world. Some are now found only in the Arctic – these are cold-climate specialists – while others can be found in the warmer regions of

southern Europe. The Lake District evidence suggests that both types lived virtually side by side in Britain all those years ago.

Using carbon-dating techniques, it's possible to determine the age of these fossils. The cold-climate beetles are 13,000 years old, but the warm-weather species are only about 50 years younger. How can beetles with such different climate preferences possibly have survived as neighbours? The only answer is that, around 13,500 years ago, climatic conditions in Britain shifted from arctic to Mediterranean extremely rapidly. The average temperature must have risen by something like 10°C (18°F) in less than 100 years, and this dramatic change had an entirely predictable effect on the ice: it began to melt.

Now ice over a mile thick doesn't melt overnight. It would have taken centuries to disappear completely, but the first trickles of meltwater must have

(ABOVE) NEWTONDALE IN THE THE NORTH YORK MOORS. TODAY THE ROUTE OF A STEAM RAILWAY, IT WAS ONCE THE DRAINAGE CHANNEL FOR WATER THAT POURED OFF THE MELTING ICE SHEETS.

(OPPOSITE) COMMON SEALS CONVERGE ON BREEDING GROUNDS AROUND OUR COASTS IN AUTUMN TO GIVE BIRTH AND RAISE THEIR YOUNG.

(PREVIOUS PAGE) THE SEVEN SISTERS IN SUSSEX. THESE TOWERING CHALK CLIFFS WERE FIRST SCULPTED BY THE RISING WATERS OF THE CHANNEL.

quickly grown into a flood of biblical proportions. The release of so much water swamped existing drainage patterns, altering the course of rivers and carving new channels in the thawing landscape.

The picturesque valley of Newtondale in North Yorkshire runs south for 6 miles (10 km) from the North York Moors. It is a huge, flat-bottomed, steep-sided valley, and yet there is only a small stream at its lowest point, narrow enough to jump across and certainly far too small to have carved a valley of this size. Appropriately, a stream like this, one out of all proportion to the valley through which it flows, is called a 'misfit'.

Around 13,000 years ago, Newtondale would have looked very different. This valley was one of the main drains for the meltwater flood, and the river flowing through it was one of the largest and most powerful ever to flow across Britain. It must have been an awesome sight as the surging flood powered its way through the hills. Loaded with huge boulders and great chunks of ice, it would have made short work of the tough grits and carboniferous limestone that make up this part of Britain, scouring them away and sweeping them down the broad valleys.

It was a similar picture across much of the country. Misfit streams, dry river channels and other flood-carved features can be found all over upland Britain, from the Gwaun Valley in Snowdonia to Seathwaite in the Lake District, from Glen More in the northern Cairngorms to Goredale Scar in Yorkshire.

The huge volume of meltwater affected virtually every drainage channel in Britain. The River Severn, Britain's longest river at 214 miles (354 km), once flowed north to reach the Irish Sea via the Dee Estuary. But during the great thaw, the retreating ice temporarily blocked this channel, forcing the water to pool up in a giant lake, Lake Lapworth. The waters finally found a southern outlet into the Bristol Channel, cutting out the Ironbridge Gorge along the way – a route that the river still follows today.

Even Old Father Thames didn't escape the effects of the great melt. Today it makes its sedate way through the orderly meadows and vales of southern England, following the course it did 13,000 years ago, but with one significant difference. Back then, this most English of rivers connected directly with the European river system – it was actually nothing more than a tributary of the River Rhine. It met the sea not at Southend but several hundred miles away, somewhere south of Lands End in Cornwall. Why? Because the lower sea level 13,000 years ago meant there was no North Sea, so the Thames simply continued flowing east into Europe until it met the Rhine flowing west. United, the rivers then turned southwest and flowed across the low-lying lands now covered by the English Channel, to finally meet the Atlantic somewhere between Cornwall and Brittany.

It took a few thousand years for the great ice sheets of the Ice Age to melt, and during that time Britain must have been very soggy indeed. Not all the meltwater drained into the sea – some collected in troughs scraped out by the ice sheets, forming many of Britain's greatest lakes: Loch Ness, Loch Rannoch, Loch Tay, Loch Lomond, and the 16 major lakes of the Lake District, including Windermere, Wast Water, Coniston and Ullswater.

(OPPOSITE) THE GREAT HORSESHOE OF CLIFFS AT MALHAM COVE IN THE YORKSHIRE DALES MUST ONCE HAVE LOOKED LIKE NIAGARA FALLS, AS A HUGE MELTWATER RIVER CASCADED 300 FT (90 M) OVER THE EDGE.

The clear, deep waters of many of these lakes are home to a fish that has a direct link with the great thaw. The arctic char belongs to the salmon family and, like the salmon, it is anadromous, which means that it starts life in freshwater streams and migrates to the ocean to fatten up, before returning to its place of birth to spawn. As its name implies, the arctic char really belongs in the chilly waters of the Arctic Ocean. But cast a line into many of the deep lakes in northern Britain and you might just haul one out, even though its preferred habitat is hundreds of miles to the north. These fish were trapped in the lakes some 13,000 years ago when the Ice Age ended, not by a physical barrier, such as a dam or landslide, but by rising water temperatures. As the climate warmed, the fish sought out the cool waters in our deepest lochs and lakes and became trapped there by the warm surface layers. They no longer migrate, even though the lakes are still linked to the ocean.

The char share many of their lochs with sea trout and salmon, which have no problem migrating through warm waters. Every autumn, rivers such as the Tay and the Tweed in Scotland, and the Shannon in Ireland, fill with migrating trout and salmon, all

(RIGHT) ULLSWATER IS A GLACIAL RIBBON LAKE 7.5 MILES (12 KM) LONG AND MORE THAN 200 FT (60 M) DEEP. ONE OF THE LARGEST LAKES IN THE LAKE DISTRICT, IT SITS IN A TROUGH SCRAPED OUT BY ANCIENT ICE SHEETS AND IS SURROUNDED BY CLASSIC ICE-SCULPTED SCENERY – U-SHAPED VALLEYS, SHARP RIDGES AND TOWERING PEAKS.

struggling upstream to reach their spawning beds. But some sea trout have taken a different evolutionary route and become permanent residents of their lakes. In Loch Ness and certain other places, these trout have evolved into monsters. When they reach about 1 ft (30 cm) in length, they switch from eating insects to preying on other trout; they become cannibals. These so-called ferox trout might grow to 18 lb (8 kg) or more – about 10 times the size of their ocean-going kin.

Some British lakes and lochs formed not from troughs gouged by glaciers but from underground cavities left by blocks of ice. Called kettle lakes, they are the end result of a long process that starts when large – sometimes very large – blocks of ice become isolated from the main ice sheet and are buried under glacial rubble. When this subterranean ice later melts, the ground above collapses and creates a bowl-shaped depression that subsequently fills with water.

Loch Leven in Kinross is one of our largest kettle lakes, measuring more than 4 miles (6 km) long and 2.5 miles (4 km) wide. It is renowned not only for having the largest concentration of breeding ducks found anywhere in Britain, but for providing a winter home to thousands of migratory ducks, geese and swans. Another kettle lake is Loch of the Lowes in Perthshire, which is famous for its resident ospreys. Ospreys became extinct in Scotland in 1916 but reappeared in the 1950s. Since 1969, a pair has been nesting at Loch of the Lowes each summer after making the long journey from their wintering grounds in sub-Saharan Africa. Today there are more than 150 pairs of ospreys in Scotland.

It wasn't just meltwater, however, that changed our islands' landscape. Like giant conveyor belts, glaciers and ice sheets pick up soil and rocks, which they drag for miles and then dump as moraines – great piles of rocky debris. When ice sheets and glaciers melt during warmer periods, their cargo is suddenly marooned. The larger boulders abandoned in this way are called erratics. Once they littered the

Ospreys became extinct in Scotland in 1916 but reappeared in the 1950s.

(ABOVE) OSPREYS ARE SPECIALIST HUNTERS THAT PLUNGE INTO LAKES AND RIVERS WITH LEGS OUTSTRETCHED TO SNATCH FISH. THEY CAN LIFT VERY LARGE FISH WITH THEIR HOOKED TALONS, AND THEIR FEET ARE COVERED WITH SHARP SCALES TO HELP THEM GRIP THE SLIPPERY PREY.

landscape, but these days most have been tidied away by farmers, builders and even landscape gardeners. Erratics are the most conspicuous type of debris left by the ice, but far greater in terms of both bulk and importance are finer materials, such as gravel, pebbles and the dust formed from rocks pulverized by repeated freezing and thawing. Much of this fine material has been spread across the British Isles by streams and rivers.

The finest debris was picked up by the wind after it dried out. The steep temperature gradient between the retreating ice and newly exposed land would have generated frequent gales, and these would have whipped up huge quantities of dust, sand and grit, creating dust storms that darkened the sky for weeks at a time. In some areas, thick layers of wind-blown dust, or 'loess', have accumulated to depths of more than 300 ft (90 m), forming the basic structure of much of the soil we have today.

(ABOVE LEFT) SANDS OF FORVIE. THIS EXPANSE OF SAND IN NORTHEAST SCOTLAND FORMED MAINLY AT THE END OF THE ICE AGE. LARGE QUANTITIES OF SEDIMENT WERE CARRIED BY RIVERS TO THE COAST AND DEPOSITED OFFSHORE. WHEN SEA LEVELS LATER ROSE, THE DEPOSITS WERE WASHED BACK ONTO THE LAND. TODAY THE SANDS STRETCH FOR MORE THAN 12 MILES (20 KM) ALONG THE ABERDEEN COAST.

(ABOVE RIGHT) THE NORBER ERRATICS IN THE YORKSHIRE DALES ARE BOULDERS DUMPED BY THE ICE SHEET THAT ONCE COVERED BRITAIN. SOME ARE AS BIG AS A FAMILY CAR. MADE OF HARD GRITSTONE, THEY WERE LEFT ON TOP OF SOFTER LIMESTONE THAT HAS SUBSEQUENTLY ERODED, LEAVING EACH ERRATIC ON A TINY PEDESTAL, LIKE A GOLF BALL ON A TEE.

The Warming Climate

It's said that the British Isles have 'no climate, only weather'. Certainly we don't suffer the extremes of summer and winter that are experienced by other countries at the same latitude. That's mainly due to one thing: a warm ocean current known as the North Atlantic Drift, which sweeps past the western coast of our islands, bringing with it a hint of the tropics. The North Atlantic Drift is an extension of the Gulf Stream, which begins in Caribbean sunshine, powers up the east coast of North America, and then crosses the Atlantic, wrapping the shores of Cornwall, Ireland and even Scotland in its warming embrace. So defined is this current that it's been described as a 'warm river' flowing across the Atlantic Ocean.

After major Atlantic storms, it's possible to find clear evidence of the North Atlantic Drift's tropical provenance. Up and down our west coast, seeds with weird and wonderful names, such as the horse-eye bean, the sea bean and the nickar nut, are occasionally washed ashore. None of the plants that produce these seeds grows wild in Britain and Ireland, but all are common in Central and South America. Take the sea bean. It's also called the monkey-ladder vine, and it thrives in the Amazon rainforest. In order for these seeds to reach the Cornish coast, they must drop into a river deep in the rainforest, float through tributaries to the Amazon River itself, pass out into the southern Atlantic, drift past Caribbean reefs and islands, and then cross the Atlantic to Cornwall – an incredible journey driven by the great current.

The impact of the North Atlantic Drift on the British Isles is profound. It's been calculated to deliver 27,000 times more heat than all Britain's power stations put together, providing about one-third as much energy to western Europe as the sun. It keeps average temperatures 10°C (18°F) higher than they

should be at this latitude. It wraps the British Isles in a warm, wet blanket, and it's the main reason why our winters are relatively mild and frost-free, and our summers warm and wet. It's something that most gardeners are deeply thankful for. In the southwest its effects are such that plants associated with much warmer climes are able to thrive. The gardens of Tresco in the Scillies and Mount Stewart in County Down, southern Ireland, are famous for their sub-tropical plant displays, where palm trees and tender exotics that are seldom found on the mainland grow well outdoors. Even more surprising are the palm trees growing 800 miles (1300 km) to the north, on the west coast of Scotland, in places such as Ullapool, Inverewe Gardens and the village of Plockton. These palms thrive at a latitude of 57°N – about the same as Churchill in northern Canada, where the sea freezes over for months at a time and polar bears wander the streets.

It's said that the British Isles have 'no climate, only weather'.

(ABOVE) THE MODERATING EFFECTS OF THE GULF STREAM CAN BE CLEARLY SEEN IN PLOCKTON ON SCOTLAND'S WEST COAST. HERE PALM TREES AND OTHER SEMI-TROPICAL PLANTS THRIVE IN THE VILLAGE GARDENS.

(OPPOSITE ABOVE) TROPICAL SEEDS FIND THEIR WAY ONTO BRITISH BEACHES COURTESY OF THE GULF STREAM.

(OPPOSITE BELOW) CHURCHILL, CANADA, IS ON THE SAME LATITUDE AS PLOCKTON BUT HERE YOU FIND POLAR BEARS RATHER THAN PALM TREES.

We haven't always reaped the benefits of a warm ocean current. During the Ice Age, the large polar ice cap blocked the current's flow, keeping Britain well and truly frozen. Only when the ice retreated could the North Atlantic Drift reach north to its present position. Its impact on the emerging British Isles was dramatic. It helped to establish the seasonal patterns that are so much a part of our landscape, and it helped encourage the next stage in the transformation of Britain.

As the ice front retreated, it left behind a battered and scoured landscape, a chaotic wasteland of rock and water that was largely devoid of life. To the south of this rubble-strewn wilderness lay the tundra of southern England and France, and beyond that was forest and grassland. These areas acted as refuges of both plants and animals that quickly spread north to recolonize Britain's newly exposed landscapes.

The first waves of colonizers were the hardiest: plants such as lichens, whose wind-blown spores can grow even on bare rock. Lichens are true wilderness pioneers, a symbiotic union between two different organisms – a fungus and an alga – that rely on each other for survival. The alga provides food for the fungus, and the fungus provides shelter for the alga, so lichens don't even need soil to get started. They grow incredibly slowly and can live for thousands of years. Some lichens still around today may have started growing not long after the ice left.

As these first plants spread and decomposed, they began to contribute organic matter to the glacial dust, slowly creating fertile soils. Other plants were quick to follow. Mosses, crowberries, bog myrtle, sedges, juniper and dwarf birch and willow all sprang up as the climate mellowed and the soil improved. These plants arrived from tundra refuges farther to the south, their seeds carried by the wind, by water and by animals (*see box* 'Seed Dispersal', p. 76). Their roots began to bind the loose glacial debris together, and as they in turn died and decayed, they added more organic material to

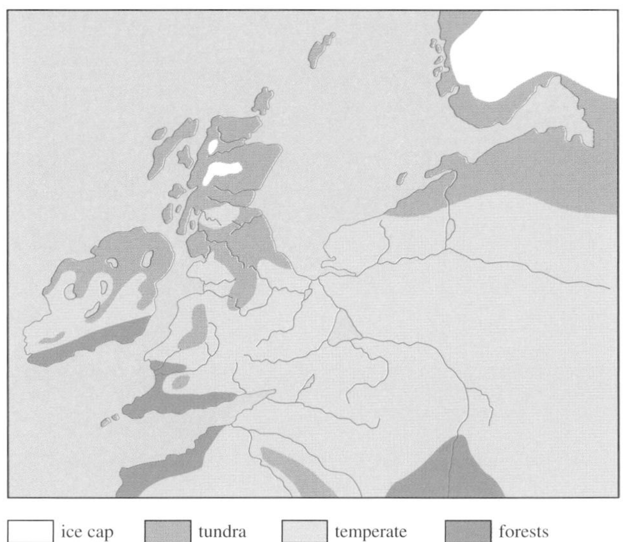

the impoverished soil. With each cycle of the seasons, the recolonization gained momentum.

The majority of the early plants were cold-weather specialists, and they can still be found growing in a few places, hanging on in only the most cold and harsh environments. Just as the arctic char descended into the depths of lochs to find cooler water, so arctic plants have had to climb mountains to find a home that is to their liking. Places such as the Burren in Ireland, the remote summits in Snowdonia, upper Teesdale in Yorkshire, Ben Lawers in Perthshire and the high plateaux of the Cairngorms are all famous for their unique alpine and arctic floras.

The arrival of lowly tundra plants clearly wasn't the end of the recolonization story. The natural cover of the British Isles is mature woodland. If we simply abandoned the land, pretty much everywhere would revert to mixed deciduous woodland in fairly short

Some lichens still around today may have started growing not long after the ice left.

(ABOVE) THE BURREN IN WESTERN IRELAND IS A PLANT-HUNTER'S DREAM. A VAST AREA OF EXPOSED AND FRACTURED LIMESTONE, THIS NATURAL 'ROCKERY' IS HOME TO MANY RARE PLANTS, SUCH AS THE SHRUBBY CINQUEFOIL.

(OPPOSITE) 11,000 YEARS AGO ONLY THE SCOTTISH HIGHLANDS STILL HAD GLACIERS. IRELAND HAD BECOME AN ISLAND AND THE ENGLISH CHANNEL WAS NEARLY FORMED.

Seed Dispersal

Just how do plants spread into new territory? They obviously can't move like animals, so to spread and colonize new habitats they must somehow broadcast their seeds, and they do this in many complex and cunning ways.

Some plants produce seeds bearing wings or parachute-like tufts so they can be blown by the wind; many of Britain's early colonizers, such as birch and pine, spread in this way. Others produce fruits that split explosively when ripe, catapulting seeds out into the surroundings.

There are hitchhikers too, dependant on animals to spread their seeds. Such plants produce seeds covered in hooks and barbs that snag the fur of passing animals, which may carry the seeds a considerable distance before dropping them or grooming them out of their fur.

Burdock is a classic British example and was the inspiration for Velcro. Mistletoe has sticky fruits that are attractive to birds, especially thrushes. The seeds stick to the bird's beak, which it later tries to clean by rubbing on bark, often on a tree some distance away. Left on the bark, the seeds grow into new mistletoe plants.

Many plants surround their seeds with a sweet, fleshy pulp – a fruit. These are irresistible to animals, which eat the fruits and so carry the seeds far and wide.

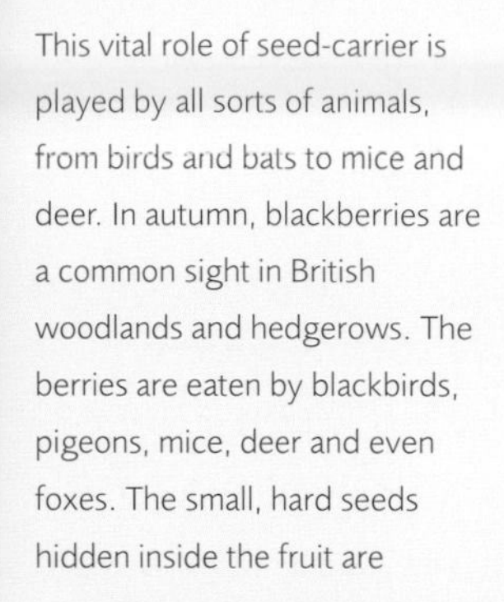

This vital role of seed-carrier is played by all sorts of animals, from birds and bats to mice and deer. In autumn, blackberries are a common sight in British woodlands and hedgerows. The berries are eaten by blackbirds, pigeons, mice, deer and even foxes. The small, hard seeds hidden inside the fruit are indigestible and pass through the intestines of these animals, to be expelled in a mound of fertilizing droppings. Some seeds dispersed in this way cannot grow unless they have passed through an animal's digestive system, where acids and enzymes trigger germination.

(TOP) POPPY SEEDS ARE SCATTERED BY THE WIND AS THE SEED HEAD DRIES OUT.
(ABOVE) THE HOOKS OF BURRS SNAG IN THE FUR OF PASSING ANIMALS.
(LEFT) DANDELION SEEDS ARE SCATTERED BY THE WIND IN DELICATE PARACHUTES.

order. And that's what happened as the ice continued to retreat and the climate warmed.

Some of the first trees to get a foothold in the new soils were birch and pine. These both prefer lower temperatures, so the post-glacial landscape was, for a while at least, perfect. Free from competitors, these trees advanced northwards at about 1300 ft (400 m) per year. At one time, virtually all of Britain was covered by them. Birch is a specialist at colonizing bare ground. It produces huge numbers of light, wind-dispersed seeds and it grows very quickly, so birch saplings are often the first trees to sprout in a forest clearing. Birches draw nutrients up into their branches and leaves each spring, before returning them to the soil surface when the leaves fall in autumn, thereby enriching their local habitat. It's estimated they can produce between 7 and 10 tons of leaf litter per acre per year.

Shedding leaves in autumn is a way of dealing with the cold winter, when trees such as birch shut down almost completely. Conifers have adopted a different strategy to survive the winter. They have reduced their leaves to narrow needles that stay on the tree all year round, allowing conifers to start producing food without delay when the warm weather returns in spring. The needles have a waxy coating that reduces moisture loss, and they contain antifreeze to protect them from the lowest temperatures.

It wasn't just plants that returned as the ice retreated. As Britain and Ireland became greener, animals spread north into the ever-changing landscape. The first to arrive were cold-loving tundra animals, such as reindeer, woolly mammoths, wild horses and Irish

SEVERAL PLANTS GROWING IN THE BRITISH ISLES HARK BACK TO OUR MUCH COLDER PAST. SPRING GENTIANS (ABOVE RIGHT) AND THE MOUNTAIN AVENS (RIGHT) ARE MORE USUALLY FOUND IN HIGH MOUNTAINS AND ARCTIC REGIONS.

elks (*see box* 'Irish Elk'). But as the forests of pine and birch spread across the countryside, these tundra animals disappeared, unable to cope with the changing climates and new landscapes. In their place came animals that were more at home in woodland.

Soon the pine forests echoed to the howling of wolves and the bugling of moose. Salmon colonized the rivers, re-establishing ancient migration patterns, and brown bears would have lined the river banks in autumn as they took advantage of the annual migration. Lynx prowled through the undergrowth, and beavers constructed lodges from mounds of sticks heaped in streams and rivers. By felling trees and building dams, they had a dramatic effect on our riverside vegetation, helping the spread of water-loving trees, such as alder.

Our largest native land mammal – the red deer – moved back into Britain around this time. We think of the red deer as an animal of open country, a symbol of the Scottish Highlands. But red deer are really woodland animals, and the Caledonian woods that sprang up after the ice melted were perfect for them. They are browsers by nature, preferring shrubs and trees to grass. They eat birch and rowan leaves, and they pull twigs, ivy and lichen from trees, especially in winter. They feed mainly in the early mornings and evenings and spend the rest of the day ruminating (chewing the cud).

As the sun grew hotter and the North Atlantic Drift extended its reach, parts of Britain grew too warm for the early colonists. Birch and conifer trees began to disappear, replaced by new arrivals from southern Europe. These new trees spread by any means available – wind, water and animal. Their varying dispersal methods, and the varying distances they had to travel, meant that different species arrived at different times (*see box* 'Seeing into the Past', p. 81).

Hazel was one of the first of the new wave to establish itself, around 9300 years ago. It was soon

(ABOVE, LEFT TO RIGHT, AND OPPOSITE ABOVE) THE WILD-WOOD WAS NOT THE INVITING PLACE OUR WOODS ARE TODAY. BROWN BEARS AND PACKS OF WOLVES WOULD HAVE BEEN COMMON; THEIR PREY WOULD HAVE INCLUDED RED DEER AND THE WILD HORSE OR TARPAN.

overtaken by a host of other trees, each spreading north as conditions became more favourable. Oaks, elms, limes, holly, ash, beech, hornbeam and maple arrived in pretty much that order. The latecomers were either warmth-loving species, such as hornbeam and maple, or just slow colonizers, such as lime.

One of the first on the scene was that most British of trees: the oak. Its seeds are spread largely by animals, especially jays. These colourful and noisy birds are prolific seed collectors, and acorns are a favourite food. They find them almost irresistible and will travel a great distance to collect them. Jays not only collect acorns but also hoard them for winter, burying each one in the ground in a different hiding place. A single jay may hoard up to 3000 acorns in one month of frenetic autumnal activity. Despite their phenomenal memories, jays occasionally forget where they've hidden some of the acorns – and from tiny acorns mighty oak trees grow.

Throughout the British Isles, the oak and all the other new arrivals battled it out in a struggle to colonize and dominate the landscape. By about 8000 years ago, our islands were cloaked in a dense forest – the wildwood had taken root.

Irish Elk

The largest deer species ever was the Irish elk (*Megaloceros*), which lived during the Ice Age. Males stood 7 ft (2.1 m) tall at the shoulder and had stupendously large antlers, in some cases measuring 13 ft (4 m) wide. Despite the name, it was neither really an elk nor exclusively Irish. It lived in Europe, North Africa and Asia, but the name 'Irish' stuck because its beautifully preserved fossils were first discovered in Irish peat bogs. 'Elk' stuck because the only deer of comparable size is the European elk (known as the moose in North America).

For a while, experts thought the Irish elk's colossal antlers were weapons, but most people now think they had a sexual function. Modern deer use their antlers in ritualized battles over mates. Males assess each other's strength by squaring up, showing off their antlers, and roaring. Usually the strongest males have the biggest antlers, so the posturing allows them to sort out their social rank without coming to blows. The highest-ranking male wins access to mates, and so passes on the genes for large antlers to the next generation.

The Irish elk disappeared from Ireland about 11,000 years ago but may have survived in continental Europe until much later. The reason for its extinction has been hotly debated. Some experts have speculated that a runaway process of evolution might have made the antlers so big that the animals could hardly lift their heads. A more likely explanation, however, is that climate change after the Ice Age led to loss of the elk's natural habitat.

(BELOW) AN IRISH ELK SKELETON

The Wildwood

So what was this great forest like? Well, for a start, it wasn't just one forest. Although virtually all of the British Isles were now covered in trees, the species varied from place to place, reflecting differences in climate, geology and soil. Five main forest provinces have been identified (*see* map, p. 83).

It's hard to imagine what this wild new forest was like. Nothing of it remains today; all our native forests have been managed and altered by people for many centuries. But there are still pockets of primeval deciduous forest left in other parts of the world, including eastern Poland and parts of North America's Appalachian Mountains, and these give us clues to Britain's appearance 7000 years ago. The ancient wildwood was probably tall and dense, packed with trees that lived to be hundreds of years old. The forest floor would have been dark and gloomy, especially in areas dominated by limes and elm, which both cast dense shade. The trees were probably closely spaced, and understorey plants were suppressed, apart from shade-tolerant species, such as ferns, dog's mercury and bramble. However, occasional sunlit clearings created by fallen trees provided a refuge for small herbaceous plants, and a flush of early flowers would have coloured the forest floor each spring.

(OPPOSITE) CALEDONIAN FOREST. THIS ANCIENT ASSEMBLAGE OF TREES – THE GNARLED SCOTS PINE, THE DELICATE BIRCH AND THE SCRUBBY JUNIPER – IS A REMNANT OF THE ANCIENT FOREST THAT ONCE COVERED THE SCOTTISH HIGHLANDS BEFORE THE ARRIVAL OF SHEEP AND IRON AXES.

Seeing into the Past

Palynology – the study of pollen – has been a vital tool in helping us to understand the changes in vegetation that took place across the British Isles after the Ice Age.

Pollen grains are extremely small – about a hundredth of a millimetre across – but they are extremely tough and can resist decay for thousands of years.

They are also very distinctive – each plant species produces a unique type of pollen, and scientists can identify many of these with the help of an electron microscope. Long after plants die, their pollen remains buried in the ground, providing botanical fingerprints that tell us which species were present in the past.

A good place to look for pollen is in the muddy floor of a pond or lake. Every spring and summer, a fresh layer of pollen sinks to the bottom of lakes and ponds and gets buried in sediment. Over years, alternating layers of pollen and silt build up, creating a record of past vegetation growing in the local area. Using a special cylindrical digging tool, scientists can remove a 'core' from the floor of a pond containing pollen deposited over thousands of years.

Studies of such pollen cores reveal that oak trees were confined to Spain, Italy and southern France during the Ice Age. Alder, elm and beech moved north and west from the Balkans, and ash trees are thought to have seen out the last cold spell in northern Italy.

(ABOVE) ELECTRON MONOGRAPH OF CHRYSANTHEMUM POLLEN

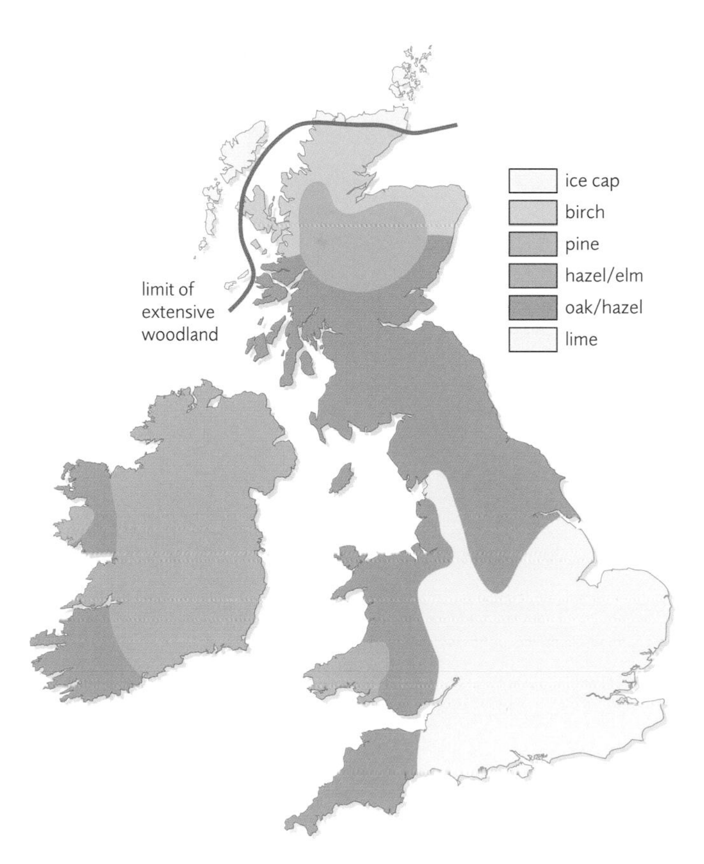

(LEFT) AROUND 7000 YEARS AGO, WOODLANDS IN THE BRITISH ISLES WERE MADE UP OF FIVE MAIN PROVINCES, WITH LIME TREES DOMINATING IN THE SOUTH, OAK AND HAZEL IN NORTHERN ENGLAND AND SOUTHERN SCOTLAND, HAZEL AND ELM IN MUCH OF IRELAND, AND PINE AND BIRCH IN CENTRAL AND NORTHERN SCOTLAND. TUNDRA VEGETATION HELD SWAY IN THE FAR NORTH, WITH TREES CONFINED TO SHELTERED VALLEYS.

(OPPOSITE) IN PLACES SUCH AS BIALOWIEZA NATIONAL PARK IN POLAND IT IS STILL POSSIBLE TO GLIMPSE JUST HOW OUR WILDWOOD MIGHT ONCE HAVE APPEARED. HUGE TREES SUCH AS THIS ANCIENT OAK WOULD HAVE DOMINATED THE FOREST AND MAY HAVE BEEN MORE THAN ONE THOUSAND YEARS OLD.

There would have been much more dead wood lying around than in our present-day forests. Fallen trees and boughs would have littered the forest floor in a dense tangle. Many of the fallen trees would still have been alive, sending up secondary shoots to create a living 'wall' and turning the forest floor into an impenetrable thicket. But not all dead trees would have fallen to the ground; standing dead trees, or snags, would have remained upright for many years, supported by their neighbours. Dead wood plays an important role in forest ecology – it is said that there is more life in a dead tree than in a living one. Standing dead trees and fallen debris create innumerable 'microhabitats' for fungi, lichens, invertebrates, mosses and birds, and they provide food for the many organisms that break down rotting wood. Nearly half of all woodland species depend on this vital resource for their survival.

> Dead wood plays an important role in forest ecology – it is said that there is more life in a dead tree than in a living one.

The wildwood did not form an unbroken canopy across the whole of the British Isles. As in any forest, there would have been openings in the canopy where rotten trees had collapsed or healthy trees had been knocked down in storms. The Great Storm of 1987, which mowed down 19 million trees in England, shows just how much devastation a storm can cause. Small gaps in the canopy are soon plugged as new trees race upwards to claim the patch of sunlight, but recent evidence suggests that extensive areas of open land may have persisted within the wildwood for long periods. Pollen studies reveal that grasses and other light-loving plants, such as willow herb and devil's bit, were constantly present in our islands. Perhaps there were meadows among the trees, kept clear by grazing animals, such as red deer and roe deer, which would have eaten saplings and kept the land open. No doubt these deer were joined by one of the true giants of the ancient woodland: the aurochs.

Now extinct, the aurochs was the wild ancestor of our domestic cattle. It was a pretty fearsome beast – not your average, cud-chewing Friesian, but a powerful and ferocious animal that stood around 6 ft 6 in (2 m) tall. Herds of these animals once roamed throughout western Asia, Europe and northern Africa. We can glimpse something of their nature in the few herds of semi-domesticated cattle that still exist in Britain, such as the famous white cattle of Chillingham.

Today none of the primeval wildwood remains. There are certainly 'ancient' forests in Britain and Ireland, but that term is applied to any land that has been continually forested for at least 400 years. That's not to say we don't still have magnificent woodlands all over the country – from the remnant Caledonian forests in Scotland to the open woodlands of the New Forest, from the mighty oaks of Sherwood Forest and Windsor Great Park to the glorious beech woods of the Chilterns. In all these places and a thousand others, it's still possible, with a little imagination, to glimpse our dim and distant past, when the woodland was truly wild.

Although the wildwood is long gone, there remain one or two trees in the British Isles that were still around when the great forest was with us. Fortingall is a small village in the heart of Perthshire, at the entrance to Glen Lyon. In the churchyard is a yew tree that was almost certainly once surrounded by wildwood. It's thought to be at least 3000 years old and possibly over 5000. It's probably the oldest living organism in Britain.

Separation from Europe

On the west coast of Wales near Borth is something quite extraordinary. Here, uncovered only at low tide and hidden amongst seaweed and tidal mud, is a whole forest of dead trees. They are all that remain of the wildwood that once flourished here. This particular forest was submerged as glacial meltwater slowly filled the low-lying basin that once connected Wales and Ireland, causing a rise in sea level of more than 300 ft (90 m).

The flooding affected the whole of Britain's coastline. Mud cores from the bottom of the North Sea show that a forest once grew there, and there are fossil forests off the shore of Skegness, in the Severn and Thames Estuaries, and in the Solent near Southampton.

(OPPOSITE LEFT) THE AUROCHS BECAME EXTINCT BECAUSE OF INTERBREEDING WITH DOMESTIC CATTLE, WHICH DILUTED THE WILD ANIMAL'S GENES. THE LAST PURE-BRED AUROCHS WAS KILLED IN POLAND IN 1627. IN 1920 TWO GERMAN BROTHERS, HEINZ AND LUTZ HECK, SET OUT TO RE-CREATE THE AUROCHS BY BREEDING DOMESTIC CATTLE WITH AUROCHS-LIKE QUALITIES. EIGHTY YEARS ON, THE RESULTS ARE IMPRESSIVE – A CREATURE WITH DISTINCTLY AUROCHS-LIKE CHARACTERISTICS, BUT A LITTLE SMALLER THAN THOSE IN THE FOSSIL RECORD. THE AUROCHS HAS RISEN AGAIN!

(OPPOSITE RIGHT) FORTINGALL YEW. TODAY THIS ANCIENT TREE SURVIVES ONLY BECAUSE OF INTENSIVE PROTECTION AND MANAGEMENT. TREE SURGEONS RECENTLY TOOK CUTTINGS FROM ITS BRANCHES TO PLANT IN FORESTS, SO THE FORTINGALL YEW WILL LIVE ON FOR PERHAPS ANOTHER FEW THOUSAND YEARS AS A CLONE, EVEN IF THE ORIGINAL EVENTUALLY SUCCUMBS TO OLD AGE.

(OVERLEAF) AT LOW TIDE, HUNDREDS OF TREE STUMPS CAN BE SEEN DOTTED ACROSS THE TIDAL FLATS OF BORTH BEACH IN WEST WALES.

Tree stumps from these ghost forests are occasionally hauled up from the bed of the North Sea by fishing boats. And as well as finding heaps of mammoth and woolly rhino bones (*see* Chapter 2, pp. 50–1), the trawlers have fished up the bones of more recent, forest dwelling animals such as aurochs, beavers, wild boar, roe deer, red deer and elks. The youngest of these have been carbon-dated as between 9300 and 8000 years old, but nothing of more recent vintage has been found. The absence of anything younger provides strong circumstantial evidence for the timing of the North Sea's formation. It's fairly safe to say that 8000 years ago we finally became the British Isles.

The severing of the land connection between Britain and Europe was a pivotal moment in our history. It suddenly became much harder for Europe's plant and animal species to reach us, and those that were already here became isolated, forming our native wildlife: about 1500 plant species (including 36 native trees), around a dozen reptile and amphibian species, and just 31 mammals. Only birds and insects were relatively free to migrate from the Continent once the seas had closed.

Although there was an extended 'window of opportunity' between the retreat of the ice and the flooding of the English Channel, not all of Europe's inhabitants managed to reach Britain. As a result, there is a clear discrepancy in the number of species found in Britain and mainland Europe. Normandy is only a few miles across the Channel, but it has 52 native mammal species, compared to our 31. Those missing from Britain's stocktake include two species of shrew, the edible dormouse, two species of vole and the beech marten.

The land link between Ireland and Wales flooded even earlier than the English Channel, so Ireland has even fewer mammals than Britain, with only 20 native species. Those that made it to Britain but not the Emerald Isle include the common shrew, the water shrew, the mole, the water vole, the field vole, the weasel and the roe deer. And it wasn't just mammals

(RIGHT) TODAY THE BRITISH ISLES ARE BLESSED WITH SOME OF THE MOST GLORIOUS COASTAL SCENERY IN THE WORLD. THE SANDS OF RHOSILLI BAY ON THE GOWER PENINSULA IN SOUTH WALES STRETCH FOR OVER 3 MILES (5 KM) AND ARE EXPOSED TO THE FULL MIGHT OF THE ATLANTIC STORMS.

that were affected – there are no snakes in Ireland. St Patrick has long taken the credit for this, but the rising sea played a far more significant role.

With the ice gone and the English Channel, Irish Sea and North Sea fully formed, the British Isles gained an extra 10,000 miles (16,000 km) or so of coastline, as well as several major islands and thousands of smaller ones. New coastal habitats were created, including cliffs, beaches, shingle banks, salt marshes, estuaries and tidal inlets. These habitats quickly filled with life. Birds and seals arrived in great numbers, establishing seasonal colonies that grew into some of Britain's most spectacular wildlife gatherings. Cut off from mainland Europe we might now be, but the variety of habitats our islands offered was expanding.

In spring and summer, the coasts of Britain and Ireland are crowded with breeding seabirds – guillemots and razorbills, kittiwakes and fulmars, puffins and petrels are all drawn to the most inaccessible cliffs and islands, where their chicks are far from the reach of land animals. The attraction is not just shelter or protection from predators – it is the abundant food to be found in the sea.

Sea animals are also drawn to this food. Migrating basking sharks arrive every spring and start swimming up the west coast, hoovering up rich 'blooms' of plankton. They are enormous fish, growing to 40 ft (12 m) long, yet they feed on microscopic prey. They swim near the sea surface with their mouths agape, using their gills to sieve plankton from the water. It has been estimated that a basking shark filters enough water in an hour to fill an Olympic swimming pool. The sharks are around the Scillies in April and reach Scotland by the end of the summer. Then they disappear to parts unknown for winter, before returning again the following spring.

Half the world's grey seals live in waters around the British Isles. Between the tides they haul them-

selves out onto rocks, sometimes in groups of several hundred. Sand eels and cod are their most important foods, but grey seals aren't fussy and take whatever fish are most abundant. In autumn they congregate in breeding colonies. The timing of births varies around the coast: pups are born in September in western Wales, October in western Scotland, and as late as November on the Farne Islands off Northumberland. The pups weigh about 30 lb (14 kg) at birth but can put on 4½lb (2 kg) a day thanks to mother's milk, which is up to 60 per cent fat – far richer than any ice cream.

Stone Age Britain

There was another species that took advantage of our rich seas. Just how successfully can be seen on the small Hebridean island of Oronsay, where the beaches are littered with shells – thousands of them. Most are limpets, but there are also cockles, mussels and the odd scallop or razor-shell. And the shells aren't just on the beach – nearby are huge mounds of limpet shells more than 30 ft (9 m) tall. Mixed among the limpets are the bones of fish, seals and whales.

These mounds are ancient rubbish tips, or middens, left by Mesolithic (Stone Age) people more than 6000 years ago. Humans had been in Britain and Ireland intermittently for hundreds of thousands of years as the Ice Age waxed and waned, and they'd disappeared more than 20,000 years ago, when the last glacial period reached a peak. When the climate warmed up, the nomadic hunters returned for good.

They survived on whatever they could find. Seafood was a mainstay on the coast, but forest-dwellers hunted deer, aurochs and wild boar. When Britain was cut off from Europe 8000 years ago, there were probably only a few thousand people here. But they soon began to have a powerful influence on the landscape. The British Isles were under siege.

Taming

the Wild

4

8000 years ago to 1700

SO FAR WE'VE TRAVELLED THROUGH ALMOST 3 BILLION YEARS OF OUR ISLANDS' HISTORY, FROM THE ANCIENT AND PRIMEVAL FORCES THAT FORMED THE BEDROCK OF THE BRITISH ISLES TO THE WATER AND ICE THAT CARVED OUT THE LANDSCAPE WE ARE FAMILIAR WITH TODAY. AFTER THE ICE RETREATED, PLANTS AND ANIMALS INVADED AND THE SEAS ROSE, CUTTING US OFF FROM EUROPE. SOON A WILD FOREST STRETCHED FROM COAST TO COAST, AND WITHIN IT

THE CYCLES OF NATURE WERE PLAYED OUT, RULED BY THE SUN AND THE MOON, THE WIND AND THE RAIN.

BUT THERE WAS ONE INHABITANT OF THE WILDWOOD THAT WAS ABOUT TO CHANGE THAT NATURAL ORDER AND LEAVE AN INDELIBLE MARK ON THE LANDSCAPE: HOMO SAPIENS. OUR ANCESTORS MIGHT NOT HAVE HAD THE SHARPEST CLAWS, THE FASTEST LEGS, OR THE MOST ACUTE SENSES, BUT THEY HAD CUNNING AND INGENUITY IN ABUNDANCE. OUR WILD LAND WAS ABOUT TO BE TAMED.

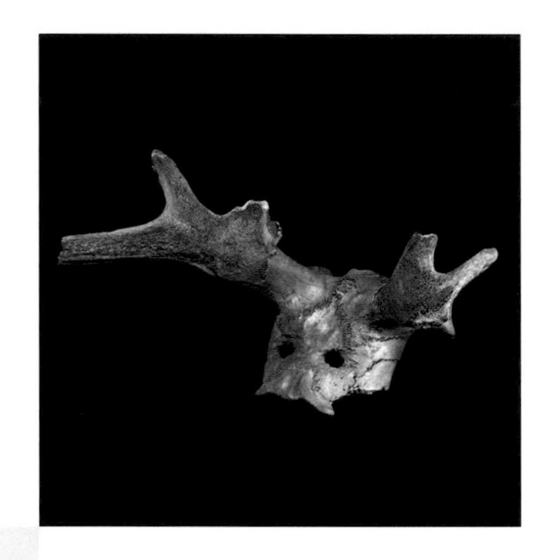

Life in the Wildwood

Imagine you're at a football match, listening to the roar of the crowd as the striker drives the ball into the back of the net. It's a first-division game, or maybe the third round of the FA Cup, with an attendance of just a few thousand. Take a careful look around the stadium and muse upon this fact: at one time the entire population of the British Isles was no greater than the number of fans at a football match. Around 8000 years ago, the average Briton probably saw fewer people in a whole lifetime than you or I see on the way to work (*see box* 'A Growing Population', p. 95).

Today there are more than 60 million people packed into these islands, and every square mile of land is owned, managed, farmed, built on, or somehow put to use to meet our need for food, shelter and entertainment. Things were very different 8000 years ago. In place of tower blocks, superstores, roads, farms and everything else we have constructed, there was nothing but wilderness. Britain was dominated by forest. Wolves, bears and lynx roamed among the trees and, from time to time, our predecessors were on their menu.

There's little doubt that our ancestors led a far lonelier existence than we do. At times they might have been the rarest creatures in Britain. They were hunter-gatherers, dotted about the great forest in small family groups and travelling widely in search of whatever food they could find. One of the best sources of evidence of how these early Britons lived is Star Carr, an archaeological dig near the seaside town of Scarborough in Yorkshire. It's not much to look at today – a field rather like any other – but beneath the surface lie some wonderful archaeological remains. They date back around 10,000 years to a period known as the Mesolithic (the Middle Stone Age).

Among the objects found at Star Carr are many sorts of tools, made not just from stone but also from

bone and timber – a clear sign of ingenuity. There are cutting tools for slicing carcasses, scrapers for cleaning hides, and barbed spears for catching fish. The strangest find of all is a set of red-deer skulls that seems to have been used as some sort of headdress. Some of the skulls, and the antlers attached to them, have been 'thinned' to make them lighter and easier to wear. It's not clear whether the headdress had a ritual purpose or was worn as a disguise when hunting. Archaeologists have also found the remains of a wooden paddle, which the inhabitants of Star Carr must have used while fishing or hunting wildfowl on a nearby lake.

We even know what Mesolithic people were eating, thanks to piles of leftovers found at Star Carr. Barbecues seem to have been popular, with wild boar, deer and aurochs high on the menu, along with a large variety of smaller animals. There are also remains of plant foods, such as hazelnuts. All in all, Star Carr

(ABOVE) AROUND 8000 YEARS AGO THE POPULATION OF BRITAIN WAS NO LARGER THAN THE CROWD AT A PRESENT-DAY FOOTBALL MATCH.

(OPPOSITE) THIS CURIOUS HEADDRESS IS JUST ONE OF MANY FINDS FROM STAR CARR IN YORKSHIRE.

(PREVIOUS PAGE) THE NEAT FIELDS OF THE YORKSHIRE DALES WOULD ONCE HAVE BEEN CLOAKED BY FOREST.

gives us clear evidence that our ancestors were hunter-gatherers, working with the seasons and living off whatever was available on the land around them.

The Mesolithic people were not entirely at the mercy of Mother Nature. Evidence from Star Carr suggests that they had already begun to shape the surrounding land. The lake that once stood at Star Carr has filled with peat over the millennia, and preserved in the peat are fine layers of charcoal. Tiny leaf fragments in these layers make it clear that the charcoal formed when reed beds around the lake caught fire, and the regularity of the layers suggests that the fires were not natural. It seems that local people were starting these fires deliberately, perhaps to improve access to the lake or to encourage fresh growth and so attract animals for hunting. Either way, fire was being used to change the landscape.

There's more evidence of this Mesolithic use of fire in upland areas of Britain, such as the North York Moors and Dartmoor. Once again, the evidence comes from layers of charcoal preserved in peat. In the uplands, fire, and the felling of the woodland that once covered these moors, have had a profound and long-lasting impact on the landscape. Rather like rainforests, hill forests soak up rain and protect the soil, helping to support a thriving ecosystem. When the trees were cleared from Dartmoor and the North York Moors, the heavy rain washed the soil into the valleys below, turning the uplands into bleak moors. Fires exacerbated the problem, changing the chemistry of the soil and creating an impermeable layer that trapped water in soggy peat bogs.

The Mesolithic hunter-gatherers undoubtedly had an impact on our landscape, but they were too few in number to threaten the dominance of the great wildwood. Their numbers were limited by the amount of food they could find, and although clever tools and the use of fire made them skilled hunter-gatherers, the population did not rise above a few thousand. But things were about to change. A revolution was heading our way, and it would transform both the landscape of the British Isles and the lives of their inhabitants forever. Once again, there is compelling evidence at Star Carr.

Among the remains of aurochs and wild boar at Star Carr are those of a much more familiar animal: the dog. It's not entirely clear where in the world the dog was first domesticated from its wild relative, the grey wolf, but the bones at Star Carr are a sure sign that dogs had arrived in Britain by the Mesolithic Period. It seems likely that they were brought from Europe and used for hunting, perhaps in a similar way to modern gundogs. Today it's the norm to see someone walking down the street with their faithful pooch, but when our ancestors first encountered tame dogs, they must have been astonished. Here was an animal that looked like a wolf, a wild and deadly

(LEFT) TAMING WILD WOLVES TO BECOME DOMESTIC DOGS WAS A PIVOTAL EVENT IN HUMAN HISTORY.

(ABOVE RIGHT) CHYSAUSTER IN CORNWALL IS ONE OF THE OLDEST VILLAGES IN BRITAIN, DATING BACK TO THE IRON AGE.

(OPPOSITE) THE CREATION OF MANY MOORLAND AREAS, SUCH AS DARTMOOR, SEEMS TO BE LINKED TO OUR ANCESTORS' USE OF FIRE.

A Growing Population

Just how do we know what the population of the British Isles was shortly after the Ice Age? Well, the honest answer is that we don't know for sure, but we can make educated guesses.

The simplest method is to estimate the amount of food a given area of woodland can provide, and the amount of food that a person needs to survive. Then we divide the first figure by the second, giving us a number of so many people per so many square miles of forest. Finally, we calculate the total area of the British Isles, giving us a very rough estimate of the whole population. Such sums suggest there were only a few thousand people living in our islands 8000 years ago.

About 6000 years ago, people began growing crops. Farmland can support far more people per square mile than can a wild forest, so we can assume that the population increased as farming spread. By 2000 years ago, there were probably hundreds of thousands of people living here.

From the Bronze Age (4000 years ago to 500 BC) there's increasing archaeological evidence of where people were living. By estimating how many settlements were in a particular area, and how many people were living in them, we can refine our calculations. We are pretty sure that by AD 1000 there were well over a million people.

In 1086 we get a helping hand from the Normans. The Domesday Book was an attempt by William the Conqueror to find out what he had acquired when he invaded England and, more importantly, whom and what he could tax. For historians it's an invaluable snapshot of life a thousand years ago. Although it covers only England – and it's a far from complete record of every-one everywhere – it suggests there were some 2 million people in Britain at the time.

From then on there's more and more well-documented evidence of how many people were living in Britain, from church records and the like. But it is not until the first full census in 1801 that we get a truly accurate figure. By that time the population had grown to well over 10 million.

enemy, yet it had been transformed into a furry friend and an obedient servant. It was arguably the first step in the taming of the wilderness and a turning point in human history, when people began to look at the natural world in a new way.

The First Fields

About 6000 years ago, the primitive boats of traders and immigrants arrived on these shores from the European mainland carrying a precious new cargo. They brought wheat, barley and – some time later – rye and oats, together with sheep, goats, cattle and pigs. All these animals had been first domesticated some 9–10,000 years ago in the Middle East, and their use quickly spread westwards to the British Isles. These days we take crops and livestock so much for granted that it's hard to imagine them not being part of the countryside. But for our ancestors farming was a novel way of doing things.

Keeping a few animals wouldn't have entailed a great change in lifestyle. Being mobile, animals could be herded from place to place by our nomadic ancestors and left to forage for themselves. The benefits were obvious: a few sheep and cows provide meat, hides and a regular supply of milk. But planting seeds and hanging around for the harvest was a much bigger change for a wandering hunter-gatherer. It meant that communities had to settle in one place for at least a year. Yet it must have had great appeal because it provided a dependable source of food. Within just a few hundred years, everybody was at it.

The effect on the landscape was immediate. New and more efficient stone tools, made by grinding or polishing rock rather than merely chipping it, had arrived with the crop plants, and they were put to good use as people began to clear the forest for farmland. We tend to think of stone tools as crude, but a stone axe, while not possessing a blade as sharp as

surgical steel, can make surprisingly short work of small trees. Experiments have shown that three men armed with stone axes can clear more than 500 square yd (400 sq m) of birch trees in just four hours. I've had a go myself, and while I certainly need a bit more practice, I can claim to have chopped down a tree reasonably quickly (though I won't be giving up my chainsaw just yet!). The arrival of the new stone tools gives this era its name: the Neolithic, or New Stone Age.

Plants and animals were quick to take advantage of the growing number of clearings in the forest. Many wildflowers prefer more open and sunny areas, and as they flourished they provided an abundance of nectar for butterflies. Grasses quickly colonized the open ground, creating substantial areas of pasture. As well as providing grazing for livestock, the expanding grasslands drew animals, such as red and roe deer, out of the forest. Where people provided the opportu-

(ABOVE) EARLY LIVESTOCK WOULD HAVE BEEN LEFT TO FORAGE IN THE WOODLAND AT FIRST – JUST AS THEY STILL DO IN THE NEW FOREST.

(OPPOSITE) ROE DEER WERE ONE OF THE MANY SPECIES TEMPTED OUT OF THE FOREST INTO THE NEW GRASSLANDS.

nities, animals and plants quickly began to colonize.

In lowland areas with poor, free-draining soil, early farmers soon exhausted the land and were forced to move on. The abandoned clearings were colonized by heather and gorse, creating the dry heathlands that still exist in places such as Dorset, Hampshire and Surrey. Although relatively few animal species can survive in these unusual habitats, those that can are among our rarest residents, and they help give heathland its unique character. Keen birders can see Dartford warblers, nightjars and woodlarks, and reptile lovers can find all six of the British reptiles: adders, grass and smooth snakes, common and sand lizards, and slow-worms.

Even the new fields of crops were a haven for wildlife. A far cry from the sterile monocultures that modern farming has produced, the fields would have been ablaze with colour in summer as wild flowers bloomed among the cereals. A bonanza of seeds, from both wild flowers and crop plants, provided food for many of our familiar garden birds, as well as for small mammals, such as voles and harvest mice. (It's around this time that harvest mice first appear in the archaeological record, so perhaps this delightful little mammal – the smallest in Britain – was carried here with the new crops.) Small rodents and birds provided easy pickings in turn for barn owls, spar-

(RIGHT) BRITAIN HAS SOME OF THE LARGEST AREAS OF LOWLAND HEATH IN EUROPE - AN IMPORTANT HABITAT INTIMATELY LINKED TO HUMAN ACTIVITY.

rowhawks and foxes, whose numbers must have quickly increased.

While the forest suffered, the creation of all the new habitats enriched the wildlife of the British Isles in many ways. But not all animals benefited from the arrival of agriculture. The predators that had once ruled the wildwood were now a threat to valuable livestock, and bear, lynx and wolf must all have been heavily persecuted by the early farmers.

Farming spread rapidly, and by around 5000 years ago the taming of our wild landscape was well underway. Perhaps as much as 10 per cent of the wildwood had been replaced with new, man-made habitats. It seems incredible that our ancestors, who still probably numbered less than 100,000 and were armed with little more than stone axes, could have had such a huge impact on the landscape. It was a revolution, and a successful one – the human population began to explode.

(ABOVE) THE BARN OWL WAS ONE OF MANY ANIMALS TO BENEFIT AS FARMLAND REPLACED FOREST.

(LEFT) LONG BEFORE WEEDKILLERS, EARLY FIELDS OF CROPS WOULD HAVE BEEN FULL OF COLOURFUL WILD FLOWERS.

(OPPOSITE) TOMBS SUCH AS WEST KENNET LONG BARROW WERE SYMBOLS OF OUR ANCESTORS' OWNERSHIP OF THE LAND.

People Power

As farming provided a more dependable source of food, the human population grew as never before. And as numbers spiralled upwards, humans became an ever more powerful force, with increasing influence on the land. That new-found strength is evident in the thousands of stone monuments that the early farmers left scattered across the countryside. Despite their great age, they remain imposing and impressive landmarks. They can be found throughout the British Isles, but the most famous are on Salisbury Plain, site of the great stone circles of Stonehenge and Avebury, as well as lesser-known monuments, such as West Kennet long barrow. West Kennet is worth particular mention as it's one of the oldest stone monuments in Britain, dating back more than 5500 years, which makes it centuries older than Stonehenge. Stones weighing many tons were dragged from miles away to be used in its construction, and thousands of tons of chalk rubble were heaped on top, creating a mound more than 300 ft (90 m) long. But why did our ancestors go to all this effort?

When archaeologists excavated West Kennet they found several stone chambers inside the mound, and within these were the remains of dozens of people – men, women and children. All the skeletons were incomplete. Skulls and various other bones were missing, perhaps taken for rituals elsewhere. Nevertheless, it seems certain that West Kennet was a tomb of some sort, a place where the local community could come to worship, or perhaps just remember, the dead. It must have been a place of enormous spiritual significance in the lives of our forebears. But monuments like West Kennet were not built just to honour the dead. The fact that they are such imposing features in the landscape suggests they had another role to play, and clues to that role lie in their location.

The land around West Kennet has always been

West Kennet is worth particular mention as it's one of the oldest stone monuments in Britain, dating back more than 5500 years.

good for farming. For the community that built the long barrow, this farmland was key to their survival. The monument served as a symbol of ownership, marking the community's family ties to the land. After all, what better way to prove your ancestral right to the land than with the bones of your ancestors? Monuments were beacons in the landscape, telling everyone that the land was taken. It's a testament to the power of these beacons that we still admire them today, even in a world of skyscrapers and tower blocks.

The purpose of stone circles is not so clear, and they almost certainly served different purposes at different times in their long history. But what all the ancient stone monuments show beyond doubt is just how far humans had come in only a few centuries. The number of people in the British Isles had increased many times over, and people were now living in much larger groups, capable of working together for a greater goal.

While our ancestors were mastering the skills necessary to construct huge stone monuments, they were also acquiring some new talents. Around 2000 BC we see the first evidence of metalwork in Britain. The ability to draw molten metal from rocks and cast it into any shape must have seemed like magic to the Stone Age farmers. Bronze, a mixture of copper and tin, was the first metal to be widely used in Britain. Easier to sharpen and shape than stone, it is much

(ABOVE) STONEHENGE STANDS AS A TESTAMENT TO THE GROWING INFLUENCE OF PEOPLE ON THE BRITISH COUNTRYSIDE.

more versatile, and bronze tools quickly became widespread. Around 1000 years later, iron began to take over. Much harder than bronze, iron was perfect for making axes to cut down trees or ploughs to work the soil. Metal tools were now the order of the day; the Stone Age was over.

Aided by new tools and technologies, people power continued to grow. Throughout the Bronze and Iron Ages the population increased, reaching hundreds of thousands, and the wildwood steadily retreated to make way for homes and farms. Crop fields, grassland and heath all expanded in the lowlands, and moorland replaced trees across Britain's uplands. The modern countryside was beginning to emerge, but there were still major changes to come – changes that had less to do with cutting down trees and more to do with imposing order on the landscape. And no one loved order more than the Romans.

(ABOVE) THIS STRETCH OF PAVING AT BLACKSTONE EDGE, ALONG THE PENNINE WAY, COULD BE PART OF A ROMAN ROAD THOUGH SOME THINK IT HAS MUCH LATER ORIGINS.

The Roman Influence

It was Julius Caesar who led the first successful Roman invasion of Britain in 55 BC, but he didn't stay long. The Romans returned home after just a few months, complaining that there was far too little silver or other booty to be found. In AD 43, the Roman emperor Claudius returned with a force of 50,000 men. Despite meeting stiff opposition, the Romans were once again victorious, and this time they didn't leave our shores so quickly.

The Roman love of order was soon imposed on Britain, and central to the new infrastructure was an efficient transport network. Roman rule depended on the swift movement of supplies and troops around the empire, and that required good roads. Of course, there were roads of a kind in Britain long before the Romans arrived, but they tended to be narrow and meandering, some survive to this day as quiet country lanes. What the Romans needed was something far more efficient, so they set about building thousands of miles of long, straight roads, carving up the British countryside.

Roman roads were so well planned that the routes stayed in use for centuries, and many survive to this day. Take the A2 between London and Dover, which follows the line of a Roman road. It's still a main trade route for people and goods arriving from continental Europe some 2000 years after it was built. Others survive only as fragments or are remembered in the names the Anglo-Saxons gave them long after the Romans left, such as Watling Street and Ermine Street. And next time you're on a shopping trip to London, imagine a Roman legion marching down Oxford Street, which was once the Roman road to Oxford.

Many of our towns also have Roman origins. Towns with names that include 'cester' or 'chester' – derived from *castrum*, the Roman word for 'fort' – are almost all Roman. As Britain prospered under Roman rule, with the population growing to more than a million, such towns became major centres of

commerce, where goods could be traded with the rest of the Roman Empire. This period of prosperity is evident in the magnificent buildings the Romans constructed here, such as the amphitheatre at Chester and the baths built over hot springs in the town of Bath.

Amid the bleak beauty of Northumberland National Park, you can see what is perhaps the most spectacular reminder of the Roman presence. Begun in AD 122, Hadrian's Wall took six years and more than a million cubic yards of stone to build. It ran for 73 miles (118 km) from Newcastle in the east to the Solway Firth in the west, and in places it was 13 ft (4 m) tall and 10 ft (3 m) thick. It was an awesome feat of engineering, surely one of the most ambitious building projects ever undertaken in Britain.

Most of what the Romans built has crumbled to ruins, but subtle reminders of their presence still live on in the countryside. High in the Yorkshire Dales are grassy meadows where dozens of different wild flowers bloom in early summer. These are hay meadows – grasslands that farmers leave to grow tall before cutting hay in late summer. Making hay was an important innovation for early farmers, helping them to feed livestock during the long British winter. And it was the Romans who introduced the tool that made it possible: the scythe. No Romans, no scythes; no scythes, no hay. Ultimately it's the Romans we have to thank for our hay meadows and all the species they support (*see box* 'Hay Meadows').

While the glorious sight of a hay meadow in full flower is worth thanking the Romans for, they made

(PREVIOUS PAGE) HADRIAN'S WALL MARKED THE NORTHERN BOUNDARY OF THE ROMAN EMPIRE.

Hay Meadows

Yellow Rattle
This plant is semi-parasitic, sucking nutrients from the roots of the grass around it. As a result, it stops grass from taking over and gives wild flowers a chance to flourish. When its seed pods are dry enough to rattle, it's said to be time to cut the hay.

Although hay meadows always make me think of my native Yorkshire Dales, they were once widespread across the British Isles. There are still a few left, but most farmers have now switched to making silage instead of hay. Silage fields are purpose-sown with grass seed, heavily fertilized (which discourages many wild flowers), and cut much earlier than hay. This means there isn't enough time for wild flowers to establish themselves and set seed, or for grassland insects and birds to complete their breeding cycles. Silage fields lack both the beauty and diversity of traditional hay meadows. In fact, hay meadows are among the most diverse wild-flower habitats in the British Isles: a single square yard of meadow might contain as many as 40 plant species. Some of the notables are listed below.

Sweet Vernal Grass
A chemical in this grass gives dried hay its characteristic sweet smell. Connoisseurs consider it the sweetest grass to chew on while contemplating the countryside!

Pignut
A member of the parsley family, this plant has 'umbels' of tiny white flowers. Its roots have small tubers (swellings) that children used to dig up and eat. Apparently they taste a bit like parsnips!

Wood Cranesbill
Another typical meadow plant, also found in hedgerows. It is a member of the geranium family but has more delicate flowers and foliage than cultivated geraniums. When the petals fall off, the remains of the flower look a little like the head and bill of a crane.

(ABOVE LEFT) VERNAL GRASS (ABOVE RIGHT) YELLOW RATTLE

another addition to the British countryside for which we owe them nothing but curses. Open your back door on a wet summer's night, switch on your torch and, if your garden's anything like mine, you'll find hordes of snails on the hunt for an easy meal. There have always been snails in Britain, but it was the Romans who introduced the ubiquitous garden snail that vandalizes our hostas and delphiniums. Native to Mediterranean lands, it was brought here as a gastronomic treat.

Ironically, the humble garden snail outlasted the mighty Romans, who withdrew in AD 410 as their empire fell apart. Other invaders took an interest in Britain in the following 500 years – including the Frisians and Saxons from Germany, and the Vikings from Scandinavia. While these new arrivals had a big influence on the people of Britain, their influence on the landscape is far less clear and there's little that suvives as a reminder of their presence.

The glorious sight of a hay meadow in full flower is worth thanking the Romans for.

(ABOVE) HAY MEADOWS ARE ONE OF OUR RICHEST HABITATS FOR WILD FLOWERS – ENOUGH TO MAKE A BOTANIST DROOL!

The Successful Snail

Land snails have long been a part of our wildlife. There are more than 80 species, and they live in a wide range of habitats, though many are just a few millimetres in size and easily overlooked. The most conspicuous of all is the garden snail, *Helix aspera*, an edible species that was brought here by the Romans and quickly adapted to its new home. It is arguably the most successful snail species in Britain, and it's little wonder why when you learn a bit about it.

Garden snails are hermaphrodites, having both male and female sex organs, and this means they can mate with any other garden snails they happen to meet. A single snail can produce 430 youngsters in a year, and seeing as they live for more than 15 years, that works out at more than 6000 babies in a lifetime. This prodigious rate of reproduction, combined with their former popularity as a delicacy, has resulted in snails ending up almost everywhere.

They are certainly a common sight in British gardens, as well as an unwelcome one – we use an incredible 400 billion slug pellets a year trying to keep them at bay. Fresh young plant growth is at the top of their menu, but snails have been known to eat paint and even brick in search of the minerals they need for a strong and healthy shell. So maybe they really are eating us out of house and home!

(ABOVE) THE HUMBLE, AND ALL TOO HUNGRY, GARDEN SNAIL.

Village England

There are parts of the British Isles that have barely changed since the departure of the Romans. Areas where the land is good for little but raising sheep, such as North Wales and the Scottish Highlands, have probably looked much the same for more than a thousand years. But in lowland areas, particularly in England, there have been major changes. The arrival of the quintessential English village, so beloved of American tourists, is one of the most obvious of these changes.

Villages had existed since long before Roman rule, as the Iron Age settlements of Chysauster and Carn Euny in Cornwall prove. However, the vast majority of English villages are medieval in origin. With their well-tended greens, duck ponds and quaint little cottages, these villages arose between the ninth and twelfth centuries. Their arrival coincides with a new method of farming that spread across a great swathe of England from Dorset to Northumbria. Historians call it 'open field' farming, and while it varied slightly from place to place, it was essentially a communal method of working the land. Because people were working together, it made sense to live together too, rather than living scattered among small hamlets, as had previously been the case. So the hamlets were abandoned and people moved into larger villages.

The green was the heart of a medieval village. Sadly, in recent times, many village greens have been lost or reduced to little more than roundabouts, but some remain. Frampton on Severn in Gloucestershire claims to have the longest green in Britain, and Wisborough Green in Sussex has a marvellous square green, flanked by horse chestnuts, that is still very much at the heart of village life. Many greens probably began life as common land on which villagers could graze a few animals; others served as the marketplace, helping the local lord to draw an income from the village. The duck ponds

that add character to so many villages also have varied origins. Some, fed by natural springs, were a source of water for villagers and their animals. Others were dug to create fish ponds, pools for soaking flax (to make linen), or reservoirs to drive a water wheel. And some were simply duck ponds, though in medieval times it was the ducks that fed us, not vice versa.

Very few of today's village buildings are as old as the villages they stand in. The original buildings, long since demolished and built over, would have been single-storey timber houses. In most villages the oldest building is the church, and a few village churches have parts dating back to the fifth or sixth century. Many more, however, were built after the next key event in the story of Britain's taming: the Norman Conquest.

The green was the heart of a medieval village.

(ABOVE) FRAMPTON ON SEVERN BOASTS ONE OF THE LONGEST VILLAGE GREENS IN ENGLAND.

1066 and All That

William the Conqueror, Duke of Normandy, invaded England in the autumn of 1066 and was crowned king of England on Christmas Day the same year. He was a ruthless leader, quelling rebellions with ferocity, while rewarding those loyal to him with huge land holdings. The Norman lords were quick to display their new-found wealth and power over both the people and the land with some of the most impressive buildings ever built in the British Isles – castles. Before the Norman Conquest there were no castles in Britain, but within a few decades there were hundreds of them, from Dover to Dublin. It's said that more stone was quarried in Britain in the century after the Conquest than was used to build the great pyramids of Egypt. Norman castles were not just symbols of power, they were sites of military and administrative rule. Many were built at high points in the land, which gave a defensive advantage as well as a more imposing appearance.

Thousands of churches and many of our most famous cathedrals date back to Norman times. With their massive masonry walls and rounded, 'Romanesque' arches, they are unmistakable. Hundreds of monasteries also appeared, and though many are now in ruins, they still have an almost magical atmosphere. In their heyday, monasteries were not just centres of worship but places of power and wealth, using the lands granted to them by Norman patrons to raise crops and sheep. Wool was becoming a valuable commodity in Europe, and the monasteries grew rich on it.

Just as the Romans did before them, the Normans made changes to the British countryside that are still with us. I went to the tiny island of Skomer off southwest Wales to find out about one of these changes. Skomer is home to substantial colonies of puffins, guillemots and razorbills, all of which are terrorized by greater black-backed gulls on the lookout for easy prey. But you have to stay overnight to see the island's real speciality.

As the moon rises, the night air fills with the ghostly calls of one of our most ungainly birds, the Manx shearwater. These seabirds are adapted to life

(ABOVE LEFT) RABBIT BURROWS ON SKOMER ISLAND PROVIDE A NEST SITE FOR PUFFINS.

(ABOVE) A THIRD OF THE WORLD'S MANX SHEARWATER POPULATION IS THOUGHT TO NEST ON SKOMER.

on the open ocean, and they're not particularly well suited to landing on terra firma – a fact that's all too clear when they crash-land by your feet. (It's because they're so clumsy that they make their arrival at night – if they came in daylight they'd be easy targets for the black-backed gulls.) Having made a functional, if inelegant, landing, they shuffle into burrows to tend their nests.

So what's this got to do with the Normans? Well, it's all to do with rabbits. Rabbits are not native to the British Isles – they were brought here by the Normans to provide food and fur. In the early days rabbits were a highly prized delicacy, so the Normans raised them on small islands such as Skomer to keep them safe from thieving eyes. Before long they escaped into the wild and started to breed ... like rabbits. And the rest, as they say, is history.

Today Skomer is riddled with so many burrows that you have to take care not to step in one and

The Norman lords were quick to display their new-found wealth and power over both the people and the land.

(ABOVE) THERE WERE NO CASTLES IN BRITAIN BEFORE THE NORMANS ARRIVED; CHEPSTOW CASTLE, OVERLOOKING THE RIVER WYE, WAS THE FIRST TO BE BUILT IN STONE.

twist your ankle. Although hazardous to us, the warrens are ready-made hotels to puffins and shearwaters. Rabbits have helped make Skomer the breeding ground for a third of the world's Manx shearwaters. And it's all thanks, indirectly, to the Normans.

As well as having a taste for rabbit pie, the Normans were keen hunters, and their favourite quarry was the fallow deer. They must have been disappointed to find none in Britain, so they brought a few with them. Like rabbits, fallow deer took to the countryside, and they are now our most widespread deer species. But even more important for the British landscape than the deer were the great tracts of forest that the Normans set aside for hunting. Today we think of forest and woodland as the same thing, but their original meanings were different. 'Forest' was a broader term, referring to any area set aside for hunting by the king or his officers, and this included woodland, heathland, grassland or even housing areas. At the height of their popularity, these hunting grounds covered at least a fifth of England. The New Forest, Sherwood Forest, Windsor Great Park and Epping Forest all owe their existence to the Norman passion for hunting.

So are these royal forests the last refuge of the wildwood? The answer, sadly, is no, for the last areas of wildwood had almost certainly vanished before the royal forests were created. Woodland did cover perhaps 20 per cent of Britain in Norman times, but it

(LEFT) WINDSOR GREAT PARK IS HOME TO SOME MAGNIFICENT OAKS, AMONG THE OLDEST TREES IN EUROPE.

was heavily influenced by people, as it had been for many centuries. Livestock wandered among the trees, trampling on and eating seedlings and preventing the understorey from regenerating. The trees themselves were carefully managed to produce poles and logs of varying sizes for making tools, fences, charcoal, buildings and ships. This was no wild woodland – it had been tamed to meet the needs of people (*see box* 'Coppicing and Pollarding').

The wildwood may be long gone, but the royal forests do hold a few treasures from the past, including some of the oldest trees in Europe. Some are old enough to have seen William the Conqueror riding past in pursuit of his beloved fallow deer. I've seen some of these trees myself at Windsor Great Park – oaks that stand like dignified giants, their rough, gnarled trunks scarred by ancient wounds, including gaping holes where they've lost limbs. These oaks are home to an extraordinary range of other organisms, among them bats, birds, butterflies, bracket fungi and some very unusual beetle species that live nowhere else. The beetles are as rare as giant pandas or tigers, though less imposing, being only a few millimetres in size.

Ironically, these trees and their many inhabitants owe their existence to people. Royal protection has doubtless helped to preserve them, but they may also have benefited from regular cycles of pollarding. Just as dead-heading sweet peas keeps them flowering, pollarding seems to keep trees sprouting for decades or even centuries longer.

The Normans played a key role in shaping the British landscape, giving us castles, royal forests, deer and rabbits. During the three centuries that followed their arrival, Britain prospered, and the population swelled to 4–6 million people as more and more land was brought under the plough. The taming of wild Britain was almost complete, but Mother Nature had a trick up her sleeve – she was about to unleash a dreadful weapon upon the people of Britain.

From Plague to Prosperity

By the fourteenth century Britain had become an important trading nation, its ports visited regularly by ships carrying goods from the European mainland. Hiding among the cargo were some unwelcome stowaways – rats – and hiding in their fur were the fleas that spread bubonic plague.

Things were already hard. The economy seems to have been in recession, and several years of bad weather had caused poor harvests and a subsequent food shortage. But it was about to get much worse. The Black Death spread like wildfire, in some places killing as many as 70 per cent of the population. In just 18 months, one in three people in Britain had died.

It was a catastrophe of unimaginable proportions. Abandoned farms went to ruin, and Mother Nature began to reclaim her land. But it was only a temporary reprieve for wild Britain, and in just a few generations the human population recovered and then surpassed pre-plague levels. So did the Black Death have a lasting impact on the British landscape?

Coppicing and Pollarding

For thousands of years people have been managing Britain's woodlands to suit their needs. The two most important types of woodland management are coppicing and pollarding. The word 'coppicing' comes from the French *couper*, which means 'to cut', and that's basically all there is to it. Trees such as willow, sweet chestnut and hazel are cut back to their base and allowed to sprout back as clusters of woody shoots, which grow remarkably quickly, sometimes by as much as 1 ft (30 cm) a week. When these reach a useable size they are harvested, repeating the cycle. In this way, woodland can be harvested over many decades and in some cases over many centuries.

Pollarding is similar, except that the tree is not cut back to ground level, but to just above head height. In former times, when livestock were left to forage in areas of woodland, this helped, to ensure that new shoots didn't all get eaten.

In a lonely valley in Yorkshire, among green fields, stands an isolated church. It's all that remains of Wharram Percy – one of thousands of medieval villages that were deserted after the plague. It used to be thought that these villages were abandoned as a

direct result of the plague, but studies at Wharram Percy have revealed a more complicated story. It seems that this village, like many others, was abandoned around 1500, long after the Black Death. The cause was a different kind of plague: a plague of four-legged monsters rather larger than rats. A plague of sheep.

Calling sheep monsters might seem to be overstating the case, but that is how they were seen at the time. In the words of medieval scholar Thomas More, 'sheep ... begin now, according to report, to be so greedy and wild that they devour human beings themselves and devastate and depopulate fields, houses

(ABOVE) THE CHURCH IS ALL THAT REMAINS OF THE ANCIENT VILLAGE OF WHARRAM PERCY.

Hedgerow Havens

A hedgerow in May, with the soft white flowers of hawthorn in full bloom, is one of the classic sights of the British countryside. And it's not just people who appreciate hedges – they provide habitats for an enormous number of species. Wild flowers, such as primroses, and wild garlic find shelter at the base of hedges, as do the food plants of many butterflies, such as Jack-by-the-hedge, on which the orange-tip butterfly lays its eggs. A single mile of hedgerow can contain more than 40 nests built by any of 60 bird species, including the long-tailed tit, whose delicate nest is a ball of lichen, moss and spider silk. An incredible 21 of the 28 lowland British mammal species breed in hedgerows, from the tiny harvest mouse to the prickly hedgehog.

(ABOVE) HEDGEROWS ARE HOME TO A RICH MIXTURE OF PLANTS AND ANIMALS.

(BELOW) THE DORMOUSE IS SURELY ONE OF THEIR CUTEST INHABITANTS!

and towns'. In the aftermath of the plague, with manpower in short supply, sheep farming became increasingly popular, partly because sheep need little looking after, and partly because their wool was still very much in demand. Landowners, keen to make a fast buck, evicted tenants and turned their lands over to sheep. It was this process that created many of England's deserted villages. (A similar process happened centuries later with the Highland Clearances in Scotland, while in Ireland it was cattle farming that drove the poor from the land.)

Wool contributed once again to a growing British economy, and those who became wealthy from it were looking for new ways to invest their profits in the land. The open field system of farming, with its complicated communal rights and obligations, didn't suit these early venture capitalists. Gradually, either by general agreement or, eventually, by an act of Parliament, the open fields were carved up into smaller, privately owned chunks of land. These new fields had to be clearly defined, and many miles of new hedgerows and stone walls were built at this time, marking the limits of ownership, just as stone monuments had done thousands of years earlier. The patchwork quilt of fields, so typical of the English countryside, was born. Even areas previously considered to be barren, such as heathlands and wetlands, came under the glare of investors as new technologies and new agricultural techniques made it possible to farm them. These few remaining wild corners of the British Isles, where nature could find refuge from human interference, were under attack.

During the sixteenth century, much of low-lying eastern England was still fenland, a wetland dominated by reeds and sedges and rich in wildlife. For the local peasants these marshes provided plants for animal fodder, roofing material, and fish and fowl to supplement their diet, but to the local landowners they were useless. To make money, the landowners would have to turn the fens into farmland, and that meant draining them. There was massive investment in a vast system of dykes and canals, built to take water away to the sea and so dry out the land. The peaty soil shrank as it dried – by yards in some places – so windmills were built to pump out the water and prevent the land from flooding again. Once there were hundreds of windmills scattered across eastern England, but most have now been replaced by more efficient, if less picturesque, diesel or electric pumps. A few windmills remain in places such as Wicken Fen and Thurne Dyke, iconic reminders of the region's past, while often only a name, such as Deeping Fen, remains of the ancient wetlands the windmills helped to drain. The fenlands were now farmlands.

As the countryside was divided by hedgerows and walls, drained with dykes and canals, and brought under the control of ploughs, axes and sheep, the wild spirit of the British Isles faded away. The wildwood was gone, replaced with a manicured landscape of fields, villages, towns and cities. Of the large predators that had ruled the forest, none remained. Lynx died out before Roman times, bears probably disappeared shortly after, and the wolf was gone by the end of the eighteenth century. All had been driven to extinction by people. In just a few thousand years, we had transformed ourselves from a handful of hunter-gatherers into millions of farmers, traders and townspeople, and in doing so we had changed the landscape of the British Isles forever. Our wild islands had been well and truly tamed.

(OPPOSITE) WINDMILLS, SUCH AS THIS ONE AT THURNE DYKE, HELPED DRAIN THE FENLANDS OF EASTERN ENGLAND, ONE OF THE LAST WILD CORNERS OF BRITAIN.

THE LION INN
TRADITIONAL BEERS . GOOD FOOD
GAMES ROOM . FAMILY ROOMS
BEER GARDENS . SHOWERS
EXTENSIVE DINING FACILITIES
FREE 24 HOUR MOORING FOR PATRONS
THE VILLAGE SHOP THURN

Forging Ahead

5

1700 to 1900

HAVING SPENT MOST OF MY OWN WORKING LIFE ON THE LAND, I ALWAYS FEEL RATHER SADDENED BY THE FACT THAT TODAY 90 PER CENT OF US WORK IN TOWNS AND CITIES. THREE HUNDRED YEARS AGO THINGS WERE COMPLETELY DIFFERENT. AGRICULTURE RULED OUR LIVES AS IT HAD DONE FOR CENTURIES, AND WE HAD CRAFTED THE LANDSCAPE OF OUR ISLANDS INTO AN ELABORATE MOSAIC OF FIELDS, WOODLANDS AND UPLAND PASTURES. TO PROSPER, MOST OF US STILL DEPENDED ON SEASONAL HARVESTS, CAREFUL

MANAGEMENT OF LIVESTOCK, AND SMALL, FAMILY-RUN BUSINESSES. BUT IN THE EARLY PART OF THE EIGHTEENTH CENTURY THIS BUCOLIC SCENE WAS ABOUT TO CHANGE.

THE KEY TO THE COMING REVOLUTION LAY UNDER OUR FEET. BRITAIN'S EVENTFUL GEOLOGICAL PAST HAD LEFT US WITH AN EXCEPTIONALLY RICH AND DIVERSE RANGE OF ROCKS AND MINERALS, AND WE NOW HAD THE TECHNOLOGY TO EXPLOIT THEM. ONE MINERAL IN PARTICULAR WAS THE DRIVING FORCE BEHIND OUR TRANSFORMATION: COAL. IT POWERED A FRENZIED ERA OF CONSTRUCTION AND INDUSTRY, DURING WHICH TOWNS AND CITIES MUSHROOMED IN SIZE AND THE COUNTRYSIDE BECAME RIDDLED WITH CANALS AND RAILWAYS, CHANGING OUR ISLANDS FOREVER.

Fuelling Change

The Yorkshire Dales still reflect what much of Britain would have looked like 300 years ago, with a scattering of remote farms and villages among a patchwork of velvety pastures and dense woods. At that time, the main export across Britain was wool, and sheep farming had shaped the land.

At the beginning of the eighteenth century, industry in Britain was minimal. The few ironworks, gunpowder manufacturers and glassworks that existed were fuelled largely by charcoal, a crude fuel produced by partially burning wood. The wood came from managed woodlands. Here trees were cut back every few years by coppicing and pollarding and then allowed to grow back in clusters of woody shoots that could be harvested for firewood ('baker's faggots'), fencing, basket-making, and charcoal. For charcoal, even-sized branches were tightly stacked and covered with turf before being set alight. Starved of air, the wood blackened and dried out without burning completely; this made a clean-burning fuel that was good for smelting iron ores. Hornbeam charcoal, in particular, was favoured by ironworkers because it produced an intense heat.

Britain had very little forest in the early 1700s. Apart from the royal forests, only managed areas of woodland were left, including the Weald, the coastal fringes of the Lake District, the Forest of Dean, and the Merthyr and Ebbw Valleys. Much of this managed woodland has now returned to the wild, and the once-coppiced trees have become multi-trunk giants that have outlived their natural life expectancy. You can find good examples just a stone's throw from London in Epping Forest. Clues to the history of such woods are also visible in surviving banks and ditches, which were built to keep deer away from newly coppiced shoots, and in the bluebells and other wild flowers that still flourish in ancient woodland.

Charcoal production was a long and laboured process, and tons of the stuff were needed for smelting iron. With the demand for iron ever increasing, desperate changes were needed. The solution came from coal.

The privatization of medieval England's open field farmland squeezed many poorer people out of villages and drove them to find accommodation elsewhere. They often ended up as squatters on less sought-after land, including Titterstone Clee in Shropshire. Here they found a new way of life, digging for coal rather than raising crops or livestock. Extracting the coal was far from easy. The miners had to hack it from the ground with crude tools and gather it by hand, often from the bottom of dangerously unstable 'bell pits', into which they were lowered by rope and bucket. Coal was considered a low-grade fuel because of its acrid, sulphurous smoke, and it was used only in glassworks or as a domestic fuel. It was

(ABOVE) 'COALBROOKDALE BY NIGHT', PAINTED BY PHILIPPE JACQUES DE LOUTHERBOURG IN 1801. TOURISTS AND ARTISTS TRAVELLED FROM FAR AND WIDE TO WITNESS THE FIERY INFERNO CREATED BY THE IRONWORKS AT COALBROOKDALE.

(OPPOSITE) THE WATER VOLE, AN ENDANGERED BRITISH SPECIES, LIVES ON THE BANKS OF CANAL WATERWAYS.

(PREVIOUS PAGE) OPENED ON NEW YEAR'S DAY 1781, THE IRON BRIDGE MARKED 70 YEARS OF INDUSTRIAL PROGRESS SINCE ABRAHAM DARBY'S PIONEERING USE OF COKE IN SMELTING. IT WAS A BOLD STATEMENT OF BRITAIN'S INDUSTRIAL PROWESS.

an inauspicious start for a substance that was to transform a nation.

Coal was no use to the iron industry at first because its impurities spoilt the smelting process. But in the early 1700s, an innovative metal-worker called Abraham Darby had the bright idea of using coke. This was a form of partially burnt coal that had lost its impurities, leaving a cleaner fuel. Coke turned out to be ideal for smelting, and the iron industry boomed. For the first time, furnaces and foundries began to congregate around sources of coal and iron, such as Coalbrookdale on the banks of the River Severn in Shropshire, where Darby set up shop. By 1750 coke was cheaper than charcoal, and coppicing went into decline. But the carefully managed woodlands still had a role to play, providing the timber props needed to support ever-deeper mine shafts and tunnels. With the development of new technologies, such as blast furnaces that could smelt iron by the ton, and new techniques, such as rolling hot iron into long rods, iron became one of Britain's principal exports.

Coalbrookdale is often described as the birth-place of the Industrial Revolution, and it's still littered with remnants of its heyday. Vast furnaces, factories and quarries lie dormant, and lumps of slag are scattered about the once-coppiced woods. Thousands of Britain's railings and cast-iron fireplaces started life here, but the most impressive relic of the industry is the Iron Bridge – the magnificent arched bridge that spans the River Severn at Ironbridge Gorge near Coalbrookdale. It was the world's first cast-iron bridge, built in 1779 by Abraham Darby's grandson (Abraham Darby III) and is still in use today.

The Iron Bridge became world famous and inspired generations of engineers to use iron as a building material, giving us yet more landmarks: the Menai Bridge, the Craigellachie Bridge, the Clifton

Black Gold

About 300 million years ago, Britain was an equatorial swamp, covered in thick forests of giant ferns, horsetails and club mosses. Dead vegetation from these prehistoric trees sank in the stagnant water, where it became preserved as peat. Compacted and heated, this eventually turned into coal. Over millions of years, as the sea level went up and down, the beds of coal were covered with layers of mud and sand washed in by river deltas, and limestone deposited by tiny sea creatures.

This geological cycle produced a layer-cake of coal, limestone and iron-rich rock – all vital raw materials for iron smelting. Their close proximity was key to the economic development of our country.

(ABOVE LEFT) THE BIG PIT (PWLL MAWR), BLAENAVON, A WORLD HERITAGE SITE. AT THE HEIGHT OF COAL PRODUCTION IN THE NINETEENTH CENTURY THERE WERE MORE THAN 160 HILLSIDE MINES AND MORE THAN 30 SHAFTS WORKING THE SEAMS OF BLAENAVON.

(OPPOSITE) SEEN FROM THE AIR, THE GROUND AT TITTERSTONE CLEE IN SHROPSHIRE LOOKS LIKE THE DIMPLED SURFACE OF A GOLF BALL. EACH DIMPLE IS THE SITE OF A SUBSIDED BELL PIT, DUG BY SOME OF BRITAIN'S FIRST COAL MINERS.

Suspension Bridge and the Forth Bridge. Iron would also be used to construct the decorative roofs of Victorian markets and railway stations, and to build the frame of the enormous Crystal Palace that once stood in Hyde Park.

New Waterways

Britain's burgeoning industries soon faced another major challenge. Bulky materials, such as coal and iron, were heavy and difficult to transport. The only means of haulage was horse and cart, and roads were either muddy and slow or expensive to use because of tolls. The answer was to transport goods by water.

Between 1760 and 1810 a network of river-fed canals was created throughout most of England, linking mines and factories to city centres. This innovation enabled horses to pull three times the amount of cargo they had been able to transport by road. The canals had a big impact on the countryside, allowing animals and plants to pass between rivers that were once separate, and carrying water to the driest parts of the country. Their banks and towpaths also provided myriad new habitats for animals and plants to colonize.

One such colonist was the water vole, which caused havoc for the canal workers, or 'navvies', by burrowing through the clay-lined banks and causing seepage. Today water voles are rare, partly because of introduced predators such as mink, and partly because of lost habitat. Luckily for them, the rebirth of disused canals as places of leisure is now helping to preserve them by keeping predators at bay.

Many of our canals are now cleaner than they used to be. Their gentle waters are a haven for carp, tench, roach, bream, perch and Britain's top freshwater predator, the pike – all of which are a lure for anglers. Canals also provide a peaceful retreat for boaters, walkers, cyclists and canoeists. It's not just animals and people that have benefited. A recent study found more than 400 species of wild flowers on a single towpath, including moisture-loving species, such as figwort, meadowsweet and hairy willow herb.

With all their locks, tunnels and boat lifts, canals are a great reminder of the engineering prowess of their constructors. Nowhere is this more apparent than at Pontcysyllte Aqueduct near Llangollen in North Wales, where you feel as if you're boating with the angels as you glide along a narrow trough of iron,

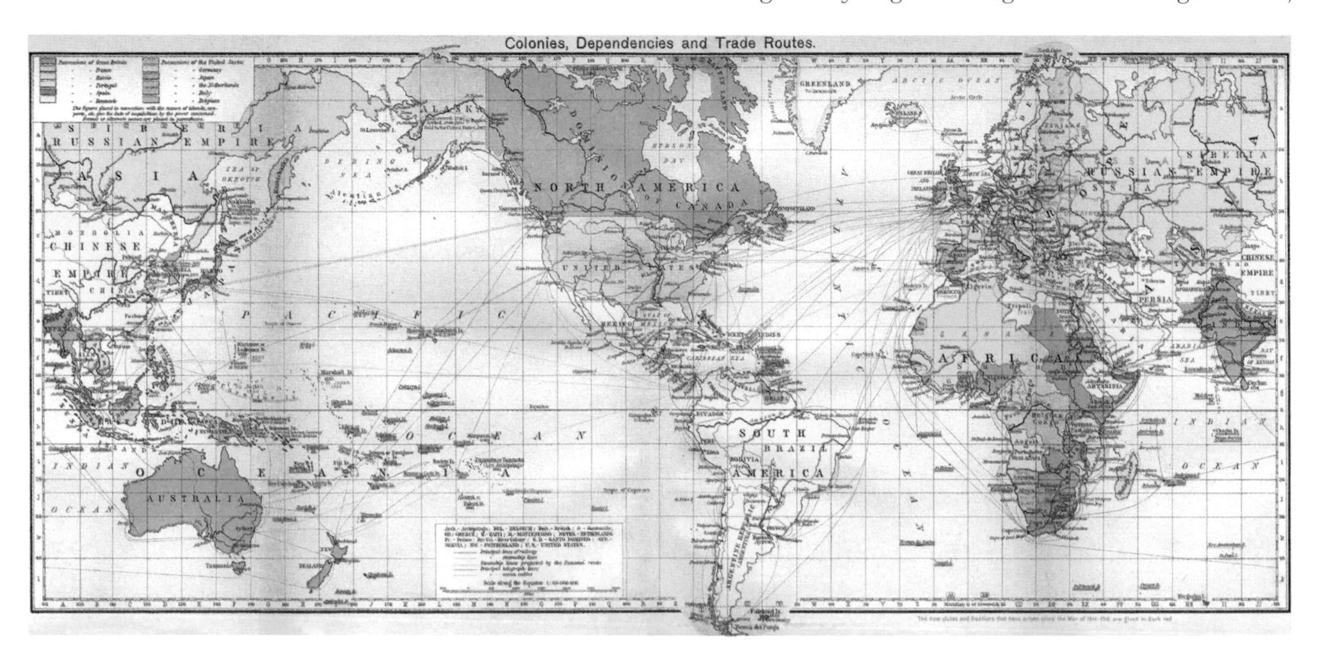

mounted on stone pillars that rise 125 ft (38 m) from the River Dee. This marvel of engineering was opened in 1805 and was originally waterproofed with a mixture of lead, Welsh flannel and boiled sugar; it's something you might want to put to the back of your mind when you're leaning on the tiller and traversing the valley below!

As well as carrying heavy materials such as coal, iron and stone, canals were used to transport fragile goods, such as porcelain, cloth and glassware, to ports for shipment overseas. During the 1700s, the value of Britain's exports multiplied eightfold as trade links were established across the world. America became the key market, but by Victorian times goods were also shipped on a grand scale to and from Africa, India, Hong Kong, Singapore, Australia and New Zealand. The trade links and shipping routes laid the foundations for the British Empire and paved the way for legions of plant collectors.

Canals are a great reminder of the engineering prowess of their constructors.

(ABOVE) PONTCYSYLLTE AQUEDUCT. MORE THAN 10,000 BOATS AND 25,000 PEDESTRIANS CROSS THIS HANDSOME AQUEDUCT EVERY YEAR. IT WAS BUILT BETWEEN 1795 AND 1805 BY ENGINEERS THOMAS TELFORD AND WILLIAM JESSOP TO LINK NORTH WALES TO THE INDUSTRIAL MIDLANDS.

(OPPOSITE) MAP SHOWING THE BRITISH EMPIRE, IN PINK, AND BRITAIN'S TRADE ROUTES.

Urban Living

There was a rural exodus during the Industrial Revolution. People abandoned the countryside and flooded into the towns in search of a wage. Many became coal miners, swelling the populations of Glasgow, Newcastle, Leeds, Sheffield, Huddersfield, Wigan, Manchester, Swansea and Cardiff. Others sought work in the new factories and mills that sprang up in the expanding towns. Fuelled by coal-fired steam engines, the textile mills, steelworks and other factories were usually built near coal mines or canals. In Stoke-on-Trent, six towns expanded and fused into one, becoming the hub of the nation's pottery industry – and all because they happened to sit on reserves of coal, clay and water. In 1800 the air here was filled with dark, sooty smoke, the product of 2000 bottle ovens at work.

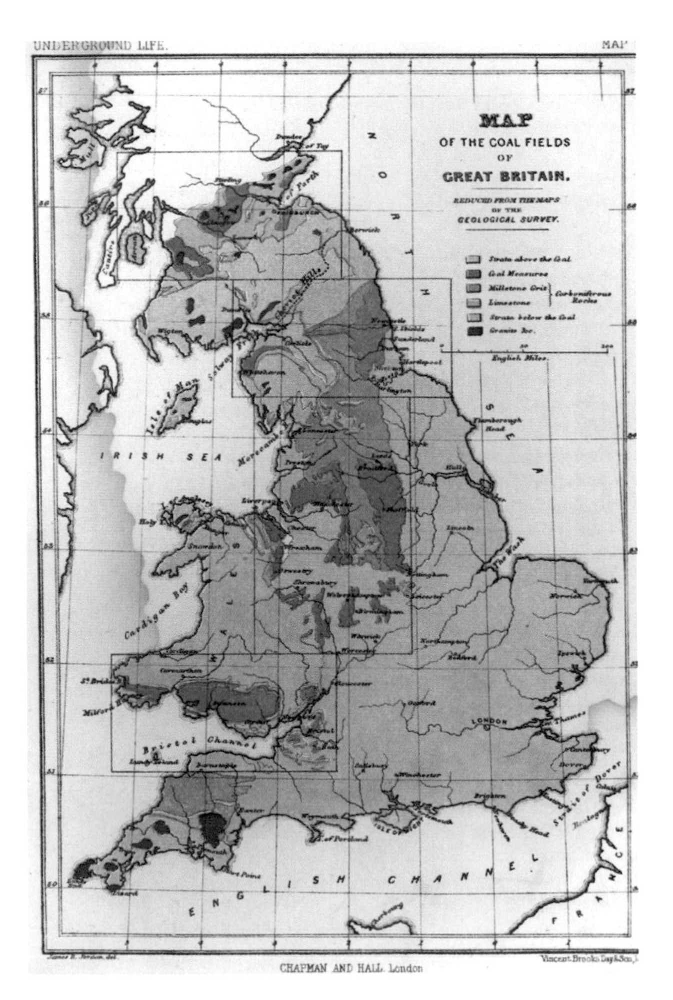

The new workers needed new homes. The construction industry boomed, and quarries multiplied as the demand for building materials shot up. The cheapest and most readily available material was clay, which was moulded and fired to make bricks. Before the canal and rail networks were established, houses had to be built from local materials, and this is

(ABOVE) HIGHGATE CEMETERY, LONDON – ONE OF THE 'MAGNIFICENT SEVEN' POSITIONED AROUND THE CAPITAL.

(LEFT) THE POSITION OF MANY OF OUR NEW AND EXPANDING INDUSTRIAL CITIES WAS DEPENDENT ON THE CLOSE ACCESS TO COAL SEAMS.

reflected in their appearance. London's clay produced yellow bricks, while Kent's iron-rich clay provided bright red ones, and Staffordshire's clay made blue bricks. To keep costs down, workers' houses were built in long terraces, with as many as possible squeezed together. Often the quality of bricks was poor, and damp seepage was common.

Slate made the best roofing material because it was waterproof and easy to split into thin slivers. By the second half of the nineteenth century, Welsh slate had become the most common roofing material in Britain. In 1873 a total of 3500 men worked at Penrhyn Quarry, one of the three main slate quarries that, between them, 'roofed the world'. Years of excavation have created spectacular slate cliffs and quarry lakes in the hills of northern Wales. Some of the quarries are still run commercially, but others are now museums or wildlife reserves (*see box* 'Quarry Havens').

By the middle of the nineteenth century, Britain's urban population had grown larger than the rural

Quarry Havens

Quarrying creates new environments and habitats, such as rocky cliffs, lakes and wind-sheltered suntraps. These can become havens for unusual wildlife. Many of Britain's oldest quarries are now SSSIs (Sites of Special Scientific Interest) for both geological and wildlife reasons. Hundreds of rare plants and insects take shelter in them, including butterflies, wasps and beetles, many listed as threatened by the International Union for the Conservation of Nature. One special habitat that's found in quarries is limestone grassland, which is home to beautiful flowers, such as pyramid orchids. Little of it survives outside quarries because of ploughing, overgrazing and neglect.

The walls of quarries form inland cliffs which provide desirable residences for some of our rare coastal birds. In North Wales the chough was a common sight from the Dyfi Estuary to the Little Orme at Llandudno until about 1865. By the end of the nineteenth century it had all but disappeared because of changes to its natural habitat. Today, it's repopulating the area by using abandoned slate quarries as nesting sites. Choughs are hard to spot, but you might be lucky to catch a glimpse of one at the Llechwedd Slate Caverns in Blaenau Ffestiniog. Another quarry resident is the peregrine falcon, which rides the updrafts over cliffs. Just watching its acrobatics and death-defying dives is enough to lift the spirits.

(TOP) PEREGRINE FALCON
(ABOVE) CHOUGH
(LEFT) BEE ORCHID

population for the first time in history, but the dramatic expansion of towns came at a price, causing overcrowding, poor housing and unsanitary conditions. Nowhere was this worse than in London. Now the largest city in the world, its air and water were filthy and stinking. The River Thames was an open sewer, and sewage was piling up in the streets and contaminating the water supplies, causing epidemics of cholera and typhoid. In one year cholera killed 14,000 Londoners, and working-class children stood a 57 per cent chance of dying before their fifth birthday, primarily because of water-borne disease. The city's graveyards struggled to cope. Bunhill, a 4-acre (1.6-hectare) burial ground just off City Road, overflowed with more than 123,000 bodies.

In 1832, Parliament organized the creation of seven new cemeteries, including Kensal Green, Highgate and Brompton, which were then on the outskirts of the city. The Victorians planted them with trees and shrubs in the hope of creating places that were pleasant to visit. Thanks partly to neglect, these cemeteries have become oases for wildlife. Nunhead Cemetery, which opened in 1840, is now a nature reserve, its crumbling church and gothic gravestones smothered with ivy and overshadowed by dense forest. Highgate is managed as a natural woodland and has been given the protection of a Grade II listed landscape.

Food and Famine

As Britain's urban population spiralled upwards, farmers turned to mechanical methods of harvesting to meet the growing need for food. Even so, England could not meet the demands of its expanding urban population, and increasing amounts of food had to be imported, much of it from Ireland.

Ireland earned the nickname 'Emerald Isle' because of its moist and often mild climate, which is the perfect environment for grass. When Britain was in the grip of the Industrial Revolution, Ireland

became a vital supplier of livestock and grain. But along Ireland's southwestern coast the drainage was poor, and the land unsuitable for almost all crops except one: the potato, a relatively recent import from the Andes mountains of South America.

In places such as the Burren in the north of County Clare, you can see just how important potatoes were to nineteenth-century Ireland. Here, parallel ridges and furrows run across endless fields, giving the ground an undulating appearance. The ridges are the remains of 'lazy beds' – raised beds about 3–4 ft (1 m) wide that were used for growing potatoes. These beds were fertilized with manure and seaweed and then built up by digging trenches between them, which improved drainage. The diggings were placed turf-down on the beds, and potatoes were pushed into the beds with a special narrow spade-like tool called a 'spud' (from which the potato is thought to have acquired its common name). Lazy beds they may have been called, but there was certainly nothing lazy about this particularly gruelling work.

A skilled potato farmer could produce 9 tons of potatoes from a single acre, enough to feed a family and its livestock for nearly a year. Unsurprisingly, potatoes became the staple crop for Ireland's tenant farmers, or 'cottiers'. Supplemented by a little milk or buttermilk (the residue of butter-making), potatoes provided nearly all the nutrients a cottier family needed to stay healthy. In fact, so nutritious was this simple, if monotonous, diet that Ireland's population exploded, growing from 3 million in 1700 to 8.5 million in 1845. By then, about 40 per cent of the population depended on the potato for sustenance – and it was about to let them down.

Most cottiers had taken to growing just one variety, a potato known as the 'lumper' because of its impressive yield. The result was an enormous monoculture of genetically similar plants, and this put the entire crop at peril – if a disease struck, the plants would all share the same susceptibility, and the infection would spread like wildfire.

Sure enough, disaster struck in 1845 when a fungal infection called potato blight (*Phytophthora infestans*) was accidentally introduced from North America, via Europe. In September of that year the cottiers watched in despair as their crops turned black and rotted. The blight returned again and again, and in six desperate years more than a million people died from starvation, and many more were ruined and forced to abandon their lodgings. To compound the problem, the Irish grain crop, which grew well though the famine years, continued to be exported to England rather than be used to feed the Irish people. Homeless and impoverished, people now turned to the government to help them find lodgings and work. There was little support for them, though some were

(LEFT) TWO GIRLS EARTHING UP POTATO RIDGES IN COUNTY ANTRIM, IRELAND.

(OPPOSITE) THE LINES RUNNING ACROSS THESE IRISH FIELDS ARE THE REMAINS OF 'LAZY BEDS' – HAND-DUG RIDGES IN WHICH POTATOES WERE PLANTED TO MAKE THE BEST OF POORLY DRAINED SOIL.

employed in workhouses and on relief roads. Road building was one of the biggest employment schemes but even this petered out due to a lack of financial support, and today you can still see the poignant remains of dead ends in places such as the Burren, near Galway. By the 1870s nearly a sixth of the Irish population had emigrated, most bound for the United States, with many dying of fever in the holds of 'coffin' ships.

The loss of people from the land has left much of Ireland free from building even today, and it is partly the wild and stunning landscape left behind that now attracts millions of visitors every year.

Landscapes and Gardens

As Britain's towns grew ever more squalid and crowded, the wealthy sought refuge in the country. Lavish country houses with extensive gardens had long been a badge of status to the upper class, and as fashions and whims changed, so did the countryside. In the seventeenth century formal gardens were all the rage. Like the gardens of French and Italian villas, they had perfectly symmetrical beds, statues, gravel paths and manicured hedges. In the eighteenth century, a more naturalistic look came into vogue. Inspired by the romantic paintings of artists such as Claude Lorraine, gardeners sought to emulate the best forms of nature and create sweeping views leading to distant ruins and temples.

At the forefront of this movement was Lancelot 'Capability' Brown, who sculpted entire landscapes to create undulating slopes, curving lakes and long vistas between trees. In keeping with Lorraine's paintings, Brown's landscapes were dotted strategically with classical Greek- and Roman-style buildings – the height of fashion at a time when the 'Grand Tour' of Europe was popular. One of his most celebrated achievements is the landscaped garden at Stowe, the Buckinghamshire estate in which he developed his style. Visiting Stowe is a bit like taking a trip back in time to ancient Greece or Rome, but in a lush green setting rather than dusty Mediterranean groves. Hidden ditches, or 'ha-has', were used to keep deer and cattle away from the grounds near the house without spoiling the view of land stretching seamlessly into the distance.

These classical garden landscapes developed at a time when industry was beginning to create blots on the landscape, and landowners were only too happy to keep progress at more than arm's length from their homes. Achieving sweeping views was no mean feat; farm buildings, woodlands and even whole villages were unceremoniously demolished in a bid to improve views. In many cases it was only the church that was left standing, provided it was considered to have 'scenic' qualities. The poor estate workers now had a far longer daily walk from their new homes.

Some eighteenth-century movers and shakers celebrated their achievements by showcasing a mill, factory or other industrial monument in their landscaped garden. The entrepreneur Ralph Allen made his bath-stone quarry the focus of his grounds at Prior Park near Bath. On the advice of Capability Brown and the poet Alexander Pope, Allen also made the most of his magnificent views over Bath and included three lakes and a beautiful Palladian bridge.

The Victorians were keen collectors of exotic plants (*see box* 'Exotic Imports', p. 134) and supplemented Britain's 36 native trees with species from all over the world. Initially, exotic trees were proudly planted as individual specimens in lawns and grassland. But by the latter part of the nineteenth century, many more species were available, and gardeners began to mix native trees with exotics in a more natural style of planting, known as woodland gardening.

Some landowners created 'pinetums' – collections of pines, firs and spruces. The industrialist and engineer Lord Armstrong filled his grounds at

Cragside in Northumberland with noble and Douglas firs from North America (known as the *tuc tuc* and *paps* to Native Americans), spruce from the Himalayas, and cypress trees from Japan. Within 30 years he had turned his barren, rocky valley into a dense woodland with more than 7 million trees and shrubs. The planting has had more than just an aesthetic effect – it has changed the valley's climate, raising the temperature by an average of 2°C (3.6°F) and increasing rainfall by about 6 in (15 cm).

Even more impressive is the exotic woodland at Inverewe Gardens in northwest Scotland, which was started in 1842 by Osgood Mackenzie, son of the local laird. Thanks to the warming effect of the Gulf Stream and a sheltering wall of pines, Mackenzie succeeded in creating a magical, subtropical forest in which Tasmanian eucalyptus trees rub shoulders with palm trees, yuccas, tree ferns and giant redwoods from California.

Inspired by the romantic paintings of artists such as Claude Lorraine, gardeners sought to emulate the best forms of nature.

(ABOVE) 'APOLLO AND THE MUSES ON MOUNT HELICON (PARNASSUS)' BY CLAUDE LORRAINE, 1680.

(OVERLEAF) BUILT IN 1738, THE PALLADIAN BRIDGE IS ONE OF OVER 30 TEMPLES AND MONUMENTS THAT GRACE THE EXTENSIVE GROUNDS OF STOWE PARK IN BUCKINGHAMSHIRE.

Gardening has been a British passion for centuries, thanks largely to our mild climate and varied geology, which allow an amazing range of plants to flourish here. When trade routes opened up between industrial Britain and the rest of the world, plant collectors set off in droves to comb the world for unusual species to show off in their arboretums and gardens. Many are now familiar garden plants, but others have escaped into the wild and become pernicious weeds.

The monkey puzzle (*Araucaria araucana*), or Chilean pine, supposedly earned its name when a nineteenth-century Englishman commented that a monkey would be puzzled to climb it. The Victorian industrialists liked its curious symmetry, which reminded them of an iron construction, and often planted it alone as a specimen tree. It was first introduced to British gardens in 1795 from its home in the mountain forests of Chile and Argentina.

Maples (*Acer*) were grown as shrubby understorey plants in sheltered woodland glades. There are around 150 Acers found throughout the northern hemisphere, but only one species – the field maple (*Acer campestre*) – is a British native; the sycamore (*Acer pseudoplatanus*) was introduced from Europe in medieval times. Maples are famed for their fiery autumn colours. Possibly the most beautiful species are the Japanese maples, many of which were brought back to British arboretums, such as Westonbirt near Tetbury in Gloucestershire. Thousands of visitors come here every year to see the autumn colours.

Magnolias, with their large, waxy, goblet-shaped flowers, were considered the most beautiful of all trees. About 120 species exist in the wild, mostly in North America, Central America and the Far East. Magnolias are of special interest to botanists, being an ancient branch of the flowering plants' family tree. They evolved before bees appeared, and their nectarless flowers are designed to be pollinated by flying beetles instead.

Most rhododendrons are native to the Himalayas and the mountains of Indochina. They were the height of fashion in the Victorian era, when every respectable country estate was planted with hundreds of them. *Rhododendron ponticum*, which probably came from Spain or Portugal, thrived exceptionally well, especially in places with acidic soil. With few natural enemies and roots that poison neighbouring plants, it spread aggressively into the countryside. When mature, it's as tall as a tree and blocks out the light, killing any plants below it. Snowdonia and other areas have been devastated by this trophy turned pest.

Buddlejas were named after the Reverend Adam Buddle, an eighteenth-century English botanist who brought the first specimen back from Africa. The most familiar species is from China: *Buddleia davidii*, which arrived here in the 1890s. Also known as the butterfly bush, it was popular with gardeners because its drooping, nectar-filled flowers are a magnet to butterflies. It turned out to be a great colonizer of wasteland and has now sown itself across the country. A frequent sight at bomb sites during World War II, it now thrives in waste ground, cracks in walls, derelict buildings and railway sidings.

Oxford ragwort (*Senecio squalidus*) gets its name from Oxford's Botanic Garden, where it was first grown in the 1700s after being brought back from the slopes of Mount Etna in Sicily. By 1794 it had spread to the city walls, but its big break didn't come until the mid-nineteenth century, when railway lines with clinker beds were laid throughout the country. The rocky, well-drained tracks were the perfect substitute for the rubbly slopes of Mount Etna and provided an easy route for the plant to spread through the country. Trains also helped, wafting the seeds along as they passed. Ragwort is now common throughout the British Isles. It is poisonous to horses but beloved of the cinnabar moth.

(ABOVE LEFT) JAPANESE MAPLE IN AUTUMN
(ABOVE CENTRE) MONKEY PUZZLE
(ABOVE RIGHT) CRAGSIDE, NORTHUMBERLAND, SURROUNDED BY CONIFERS AND RHODODENDRONS

Steaming Ahead

Despite the extensive canal system and many new roads, Britain lacked a fast and reliable means of public transport in the early nineteenth century. So, between 1830 and 1860, the Victorians set about building an extensive railway network. Steam locomotives were developed from pumping engines used to drain water from coal and tin mines. They powered trains on tracks inspired by those used to haul coal carts around mines. These first trains must have seemed amazingly fast and futuristic to the Victorians.

The railways soon made their mark on the landscape. They freed industry from the canals, allowing factories to spread to every corner of Britain, from Cornwall to Scotland. Cuttings, tunnels, bridges and embankments carved their way through the countryside, and acts of Parliament granted the railway companies the right to lay lines straight through private estates and even ruined castles! In urban

(ABOVE) GLENFINNAN VIADUCT, BUILT BETWEEN 1897 AND 1901, FORMS PART OF SCOTLAND'S WEST HIGHLAND LINE, WHICH CARRIES TRAINS BETWEEN FORT WILLIAM AND MALLAIG. IT WAS THE FIRST CONCRETE VIADUCT IN THE WORLD.

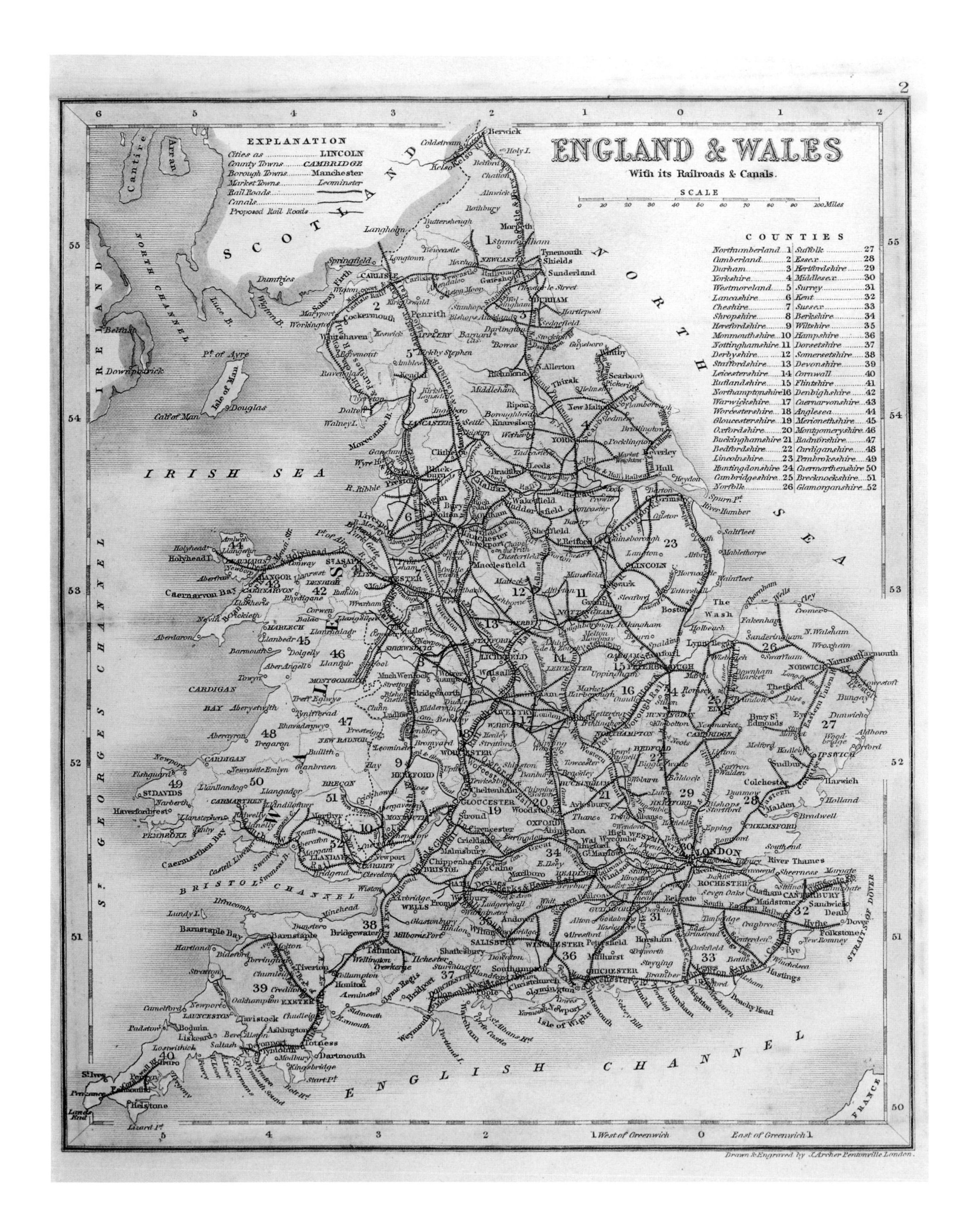
ENGLAND & WALES
With its Railroads & Canals.
SCALE
0 10 20 30 40 50 60 70 80 90 100 Miles
EXPLANATION
Cities as .... LINCOLN
County Towns .... CAMBRIDGE
Borough Towns .... Manchester
Market Towns .... Leominster
Rail Roads
Canals
Proposed Rail Roads
COUNTIES
Northumberland 1
Cumberland 2
Durham 3
Yorkshire 4
Westmoreland 5
Lancashire 6
Cheshire 7
Shropshire 8
Herefordshire 9
Monmouthshire 10
Nottinghamshire 11
Derbyshire 12
Staffordshire 13
Leicestershire 14
Rutlandshire 15
Northamptonshire 16
Warwickshire 17
Worcestershire 18
Gloucestershire 19
Oxfordshire 20
Buckinghamshire 21
Bedfordshire 22
Lincolnshire 23
Huntingdonshire 24
Cambridgeshire 25
Norfolk 26
Suffolk 27
Essex 28
Hertfordshire 29
Middlesex 30
Surrey 31
Kent 32
Sussex 33
Berkshire 34
Wiltshire 35
Hampshire 36
Dorsetshire 37
Somersetshire 38
Devonshire 39
Cornwall 40
Flintshire 41
Denbighshire 42
Caernarvonshire 43
Anglesea 44
Merionethshire 45
Montgomeryshire 46
Radnorshire 47
Cardiganshire 48
Pembrokeshire 49
Caermarthenshire 50
Brecknockshire 51
Glamorganshire 52
SCOTLAND
NORTH SEA
IRISH SEA
NORTH CHANNEL
ST. GEORGES CHANNEL
BRISTOL CHANNEL
ENGLISH CHANNEL
IRELAND
WALES
FRANCE
LONDON
West of Greenwich
East of Greenwich
Drawn & Engraved by J. Archer Pentonville London.

areas, whole streets were knocked down to make way for trains. New towns, such as Crewe, Ashford and Swindon, appeared, providing housing for the men who built the locomotives and carriages.

With the arrival of the train, building materials could be transported from far and wide. Now the Victorians could mix stones and bricks of different colours for decorative effect. One of the most ornate of the new multicoloured buildings was the Natural History Museum in London's South Kensington.

But the railway's greatest effect on the landscape came as a result of the new lifestyle opportunities it brought. People of all backgrounds and classes could now travel across the country in a matter of hours. There was a surge of interest in the countryside, and seaside resorts sprang up from Blackpool and Bournemouth to Bognor Regis and Scarborough.

With her Highland home at Balmoral, Queen Victoria made it fashionable to visit Scotland, now only a few hours from London. Everyone who was anyone followed, buying up private estates throughout the north. Salmon fishing and deer hunting were the rage, but red grouse became the mostly highly prized game, famed for being a very tricky, but tasty, target. England's upper crust took the train to Scotland every August to enjoy the fresh air and the sport. Even the Houses of Parliament went into recess so that ministers could head north for the 'Glorious Twelfth' of August – the start of the grouse season.

Hunting Grounds

In the eighteenth century, much of Britain's farmland had been divided by hedges into small fields. Quite by chance, it was the perfect habitat for game birds, the hedges providing the cover that these ground-dwellers naturally seek. Birds such as pheasant and partridge soon became a more popular target for hunters than deer, a trend that was helped by improvements in guns and the introduction of the pointer (a breed of dog that points to the bird with its

(ABOVE) HEATHER GROWING IN THE CAIRNGORMS. THIS PATTERN IS PRODUCED OVER THE YEARS BY REGULAR, CONTROLLED HEATHER BURNING IN THE SPRING/WINTER, A PROCESS KNOWN AS MUIR BURN.

(OPPOSITE) MAP OF THE RAILWAY NETWORK IN ENGLAND AND WALES, C. 1850.

body and tail in a perfectly straight line). By the Victorian era, hunting game birds was a national obsession. There was nothing the Victorians liked better than a good shooting party, and the size of their 'bags' was paramount; a visiting maharajah once shot 780 partridge in a single day in Suffolk.

The Industrial Revolution and the British Empire generated untold wealth, and much of it was spent on land. By 1874, estates of more than 10,000 acres accounted for a quarter of Britain. Landowners expanded their estates by buying the surrounding farms, which were often then turned into shooting grounds. Gamekeepers were employed to rear and protect the birds and manage the land to maximize their numbers. In 1871 there were some 17,000 gamekeepers in Britain, and they had a huge effect on both the countryside and the wildlife. A key part of their job was predator control – anything that might threaten game birds was shot, poisoned or trapped. In Scotland, ospreys and white-tailed sea eagles became extinct, and wild cats disappeared from all but the most remote places. In the south, red kites were wiped out everywhere except central Wales, and buzzards were eradicated in eastern England. Thankfully, red kites, ospreys and sea eagles are now returning to their former haunts.

Although the game reserves were bad news for Britain's carnivores, they did have some beneficial effects on the countryside. Estate owners spent a lot of money creating the perfect environment for their birds, and that environment provided habitats for other species. In the lowlands, woods were thinned to encourage the growth of shrubs, providing ground cover for the birds. Hedgerows were maintained and coppiced woods were planted. The ideal planting was a mixture of oak, beech and sweet chestnut, with a few conifers for roosting. These open woodlands were ideal for wild flowers, and the flowers in turn provided food for woodland butterflies that struggle to survive in denser forests managed for timber.

Game Birds

The common pheasant (*Phasianus colchicus*) is the most numerous game bird in Britain. Native to the Far East, it was imported many hundreds of years ago and is now a common sight in country lanes, fields and woodlands. Millions are reared and released each year for shooting.

The red grouse (*Lagopus lagopus*) lives in the heather moorlands of Scotland, Ireland and northern England and is highly prized, both by hunters and chefs. Its high value has helped to conserve its habitat for more than 200 years – in areas where shooting continues, only 17 per cent of the heather habitat has been lost, compared to 50 per cent in areas where shooting has stopped.

The grey partridge (*Perdix perdix*) was once shot for sport but is now under threat. Its presence is seen by conservationists as a sign of an area's ecological health. Its decline seems to be linked to lack of food for newly hatched chicks. The Game Conservancy Trust is now encouraging farmers to put 'beetle banks' in their land to help the partridge and other wild birds to recover.

(TOP) PHEASANT
(ABOVE) GREY PARTRIDGE
(LEFT) RED GROUSE

(OVERLEAF) LONDON PLANE TREES IN RUSSELL SQUARE

Long ago, the uplands of northern Britain were covered with forests of conifer and birch, but after centuries of sheep grazing, heather came to dominate the landscape. Regular burning of the heather was first practised to improve grazing for sheep, but later it was adopted to encourage grouse. Fire suppresses bracken and so preserves the heather, and it creates a patchwork of heather stands of differing ages. Red grouse like this as they nest in the older heather, which provides the best cover, but they prefer to feed in recently burnt areas, where there is more new growth. Heather burning is still practised to encourage grouse, and it also appears to benefit mountain hares, curlews and snipe, among other species.

Park Life

While the wealthy took advantage of the countryside, the closest the urban labouring class often got to nature was in city gardens. The great and the good increasingly felt that healthier conditions were needed to support a strong workforce, so efforts were made to improve the urban environment.

Municipal parks were opened in Preston, Derby, Manchester, Glasgow and Southampton, making them more pleasant places to live. The first municipal park was in Birkenhead, opposite bustling Liverpool. It was laid out in 1843–7 by Joseph Paxton – the brains behind the Crystal Palace – and included landscaped gardens, artificial lakes, and cricket and football pitches. It's interesting to reflect that this 226-acre (90-hectare) design, which you can still enjoy today, was copied by New York in the form of Central Park. I wonder if New Yorkers know just what they owe to Birkenhead?

The lot of the urban dweller improved apace now. Two acts of Parliament encouraged donations of land for conversion into parks; local authorities were given powers to protect land for recreation; and organizations such as the Commons Preservation Society fought to preserve what was formerly common land. By the end of the nineteenth century there had been a sea change in attitudes to the environment, and green spaces were seen as places to be conserved rather than developed. This sentiment gave birth to conservation organizations such as the National Trust, founded in 1895.

London's filthy air posed a problem for plants. Polluted by factories and thousands of coal fires, it was sometimes so thick with soot that sunlight struggled to reach the ground. What the Victorians needed was a tough tree that could thrive in such conditions, and they found it in the smog-resistant London plane (*Platanus* x *hispanica*). Despite its common name, this tree is not a true Cockney. It's an introduced hybrid, a cross between two species from opposite sides of the world: the American plane from the USA and the Oriental plane from southeast Europe, both of which were imported to Britain as ornamentals.

Plane trees planted 200 years ago are still growing vigorously in London's squares. They may yet become the largest trees in southern Britain – they are thought to live for up to 500 years and can grow to 100 ft (30 m) tall. In spring the young leaves are covered with fine hairs to protect them from the sun; these trap dirt particles in the air. The hairs then drop to the ground with the pollutants, revealing shiny green leaves that are easily washed clean of soot and dirt by rain. The plane also sheds its bark, preventing breathing pores from becoming clogged with dirt. It's a bit like a face pack really! In this way these trees can remove an amazing 85 per cent of grime from the surrounding air, making city life more pleasant.

The dramatic changes that have affected Britain over this 200-year period are mind-boggling. They have had such far-reaching consequences that they still touch our lives today. They have influenced what we do, where we live, and what we see – from a train window, a canal bank, down a street or out on the pier. The industrial era may be seen by some as a dark period in our history, but it was also an enlightening period, which has given us much of our most treasured heritage and countryside today.

The Modern Era

1900 to the present day

YOU DON'T NEED YOUR GRANNY TO TELL YOU THAT THE PAST CENTURY HAS SEEN THE BRITISH ISLES CHANGING FASTER THAN AT ANY OTHER TIME IN HISTORY. BY 1900 THE INDUSTRIAL REVOLUTION HAD LEFT BRITAIN WITH AN URBAN ENVIRONMENT MUCH LIKE TODAY'S, BUT THE LANDSCAPE WAS NOT ONE WE WOULD INSTANTLY RECOGNIZE. HORSE-PLOUGHED FIELDS WERE BORDERED BY WOODLANDS AND CRISSCROSSED BY CART TRACKS

AND STREAMS. MUCH OF OUR COASTLINE WAS MARSHY, GRAZED BY THE HARDY LOCAL LIVESTOCK, AND THE UPLANDS WERE WILD AND OPEN. THE PACE OF LIFE WAS SLOWER THAN TODAY, WITH MOST TRANSPORT STILL HORSE-DRAWN AND MOST HOMES WITHOUT ELECTRICITY. IT'S A PERIOD THAT MANY OF US THINK OF NOSTALGICALLY, A WORLD AWAY FROM TODAY'S MOTORWAYS AND HIGH-SPEED TECHNOLOGY.

So what were the key events that helped to create the modern landscape we know today? One important factor was war. The food shortages of the war years were catalysts for change, kick-starting the drive to make Britain self-sufficient in food. The rise of machines helped make farming more efficient as horse-drawn carts gave way to tractors. In the towns and cities car usage exploded, and new roads drove straight lines through the countryside. But there was also a growing appreciation of our beautiful landscapes as we sought to restore the balance of nature. One good example of this is to be found in Britain's forests.

Foreign Forests

At the beginning of the twentieth century, Britain's woodlands were in crisis. Only 5 per cent of the land was forested, and the diminished woodlands were mainly in private hands, used by wealthy landowners for pheasant shooting and deer stalking, or simply left alone. World War I changed all that. Timber was urgently required to build trenches and to provide pit props for the coal mines that fuelled so much of our arms industry. Half a million acres of forest were felled, mostly from private estates, but it wasn't enough – timber was in desperately short supply at a time when it couldn't be imported. Drastic action was necessary, so after the war the government set up the Forestry Commission, which had the sole purpose of increasing timber stocks; in 1919 the first Forestry Commission trees were planted.

Trees take time to grow, time that we couldn't afford. So Britain's best foresters were given the task of finding species that would grow fast in our wet climate and thrive on even the poorest soils. North American conifers provided the perfect solution, along with a handful of other exotic species. The sitka spruce, named after its home town in Alaska, quickly

became the forester's favourite, growing three times faster than the native oak and taking only 40 years to mature. Soon a whole host of exotic trees, such as Douglas fir, Japanese larch, lodgepole and Corsican pine, were springing up around the countryside.

It wasn't just the tree species that were alien to Britain, it was also the way they were planted. They grew in precisely straight lines, so close that light could hardly penetrate the canopy, and the edges of forests were abrupt and unnaturally straight. Britain's uplands began to look like a giant chequerboard as blocks of forest marched across the land. Although ugly, these new conifer forests were a vital resource during World War II, providing 50 million cubic ft (1.4 million cubic m) of wood and creating jobs for thousands of people.

Scotland supported the largest plantations, and in the postwar years its forested land coverage greatly increased. Large-scale plantations also appeared in

(ABOVE) IN TRUE WAR-TIME SPIRIT, THE FORESTRY COMMISSION EMPLOYED A SPECIAL WOMEN'S TIMBER CORPS, FONDLY NICKNAMED THE LUMBERJILLS.

(OPPOSITE) FARMLAND COVERS THREE-QUARTERS OF BRITAIN'S LANDSCAPE TODAY.

(PREVIOUS PAGE) COMBINE HARVESTERS NEED VAST, OPEN AREAS OF AT LEAST 100 ACRES (40 HECTARES) TO WORK EFFICIENTLY. THIS HAS LED TO THE LOSS OF MILES OF HEDGEROWS.

southern Wales, Breckland in Norfolk, and the Cheviot Hills of Scotland and Northumberland. Kielder in Northumberland became the largest man-made forest in Europe. Planting on inaccessible upland areas such as Kielder became easier with the mechanical revolution of the 1950s and 1960s, when tractors replaced horses and chainsaws replaced axes.

The new forests have been a mixed blessing for wildlife. Birds that live in moorland and open hills – such as golden plovers, dunlins, curlews, skylarks and meadow pipits – have seen their natural habitat dwindle, and their numbers have fallen seriously. Plantations can't support as many species as natural woodland because they're just too tidy, too much of a monoculture. The densely packed trees, all the same age and the same species, allow so little light to reach the forest floor that only toadstools, lichens and the occasional fern can grow underneath. There are no dead trees or rotting logs for beetles and other small

(ABOVE) DESPITE THE NEW TECHNOLOGY, DRAINING AND PLANTING BRITAIN'S UPLANDS WASN'T ALWAYS EASY, AND MANY TRACTOR DRIVERS GOT BOGGED DOWN.

(OPPOSITE) OUR NATIVE RED SQUIRREL IS BETTER ADAPTED TO CONIFER PLANTATIONS THAN THE GREY SQUIRREL, A RECENT INVADER FROM NORTH AMERICA.

creatures to live in, no hollows for birds and bats to roost in, and few flowers for insects.

But, rather surprisingly, modern plantations aren't quite the wildlife deserts you might assume. Well-managed forests contain blocks of trees at different stages of growth, creating a patchwork of habitats for all sorts of animals. A young plantation is often fenced off to keep deer away from the saplings, and this encourages a thick sward of grasses. These are colonized by animals such as short-tailed field voles, which, in turn, attract birds such as the short-eared owl. Shrews and foxes enjoy rooting around in the tussocky grass for spiders and beetles, as do insect-eating birds, such as the grasshopper warbler. Once the conifers form dense thickets at around 12 years old, such food becomes scarce, but the trees now provide nesting sites for wrens, willow warblers and hedge sparrows. When the trees are fully mature and the undergrowth disappears, the balance of species shifts again. Now the forests are valuable breeding sites for goldcrests and important refuges for our native red squirrel. Conifer plantations are the one place where they seem to hold their own against their American cousin, the grey squirrel.

Native to eastern North America, grey squirrels were brought to Britain by the Victorians as novelty animals. They were released at about 30 sites around the country, starting with Cheshire's Henbury Park in 1876, and they quickly spread into the wild. Their proliferation has coincided with the rapid demise of our native red squirrels, which seem unable to survive in the presence of greys. The reasons for this are not fully understood. The key may be the greys' ability to outcompete the red squirrels for food, especially in deciduous forests. But in mature conifer plantations, such as the Formby Point Reserve in Merseyside, red squirrels are thriving, and as this National Trust reserve is open to the public, the squirrels have become quite tame. With a little patience you can enjoy a wonderful wildlife encounter.

Britain's woodland has more than doubled in area since the start of the twentieth century, and much of this increase is due to new plantations. Although the Forestry Commission started out as a timber producer, the emphasis has now shifted towards multi-purpose forestry, where conservation, recreation and productivity are equally important. Nowadays plantations often contain natural-looking clumps of native broad-leaved trees, which are far easier on the eye, and instead of being planted in square blocks, plantations follow the natural contours of the land. But one thing is certain: the highs and lows of the twentieth century have changed our forests profoundly. In 1924 about two-thirds of Britain's forest consisted of broad-leaved trees, but today that same amount is made up of conifers, about half of which are Alaskan sitka spruce. Our woodland is now considerably more cosmopolitan than at the start of the century.

If you're a regular walker in the countryside, you'd be unlikely to say that woodland dominates our landscape – and it hasn't done for thousands of years. The thing you notice more than anything else in the countryside is farmland. It takes up three-quarters of Britain, and anything that affects farming has a huge impact on both our landscape and wildlife.

Farming Revolution

Farmland is all around us. On a country drive you're likely to pass through mile after mile of rolling wheat fields, pastures full of cattle and sheep, or well-stocked fruit orchards. Depending on where you are in the country, the agricultural mix will be slightly different. In the wet and hilly west, you might be surrounded by lush green pasture and livestock; in the flat and fertile plains of the east, golden cereal fields are more likely to make up the scenery. But it hasn't always looked the way it does today. Agriculture has experienced enormous highs and lows in the last 100 years, and every blip has had an effect on the landscape.

It's tempting to think the countryside was a rural idyll at the start of the twentieth century, a hotchpotch of small farms in harmony with nature, their small fields divided by thick hedgerows, horses working the land, and songbirds everywhere. But the reality of life for farmers was far from rosy. Farming was in deep depression and many farms were neglected. Hedges were overgrown, drains were clogged and farmers were short of money to fix anything. The land might have been a wildlife haven, but it certainly wasn't in a healthy state for agriculture.

Before World War I, we imported much of our food instead of producing it. We ate wheat from North America, beef from South America, and lamb from New Zealand and Australia. Cheese travelled all the way from Canada, bacon came from Denmark, and eggs and fruit came from all over Europe. In fact, almost anything edible could be imported cheaply. But during the war, our trade links were jeopardized, which put us at great risk of a food shortage. The government had to intervene. It offered farmers generous incentives to plough their neglected fields and switch to high-calorie crops, such as potatoes, cereals and sugar beet.

By the time World War II began, farming was profitable again. This meant money in the bank for the farmers, something they hadn't experienced for a while, and that money could be used for capital investment – more often than not for glossy new farm machinery.

Before World War II there were only 150 combine harvesters in the entire country, but by 1943 that number had increased tenfold. The number of tractors doubled, and suddenly the transformation of land, encouraged by the government, became quicker and easier. During and after the war ploughing became a national sport as grants were handed out for turning pasture, heath and moorland into arable land to grow ever more food. It was a time of rapid change, and wildlife struggled to adapt.

One casualty was the black grouse, which used to be common throughout southern and central England. Its range is now restricted to a few areas in

It's tempting to think the countryside was a rural idyll at the start of the twentieth century.

(ABOVE) THE NUMBER OF COMBINE HARVESTERS IN BRITAIN SHOT UP DURING WORLD WAR II.

(OPPOSITE) IN THE EARLY YEARS OF THE TWENTIETH CENTURY FARMING WAS STILL LABOUR-INTENSIVE, SOMETHING THAT WOULD CHANGE AS MACHINERY BECAME MORE ADVANCED AND WIDESPREAD.

All this intensification had an effect on the landscape. What was once a mosaic of higgledy-piggledy fields, with overgrown hedges, pockets of woodland and damp marshy areas, became homogenized and regimented. Small fields were combined to make huge prairies, as all obstacles were cleared to make way for the mighty combine harvester.

Larger fields meant fewer hedgerows. Many had already disappeared in the interwar years because of urban sprawl, but during the sixties hedgerows were cleared at an alarming rate. Between 1945 and 1970 we lost about a quarter of our hedges, at an average rate of 5000 miles (8000 km) per year. Their decline has gone hand in hand with the decline of many of the bird species that rely on hedges for nesting, feeding or cover. Just a mile of mixed hawthorn and elm hedge can support more than 40 pairs of nesting birds, so the loss of this important habitat has a heavy toll.

Hedges have been a part of our landscape for hundreds, perhaps even thousands, of years. Though many were planted after the Enclosure Acts of the eighteenth and nineteenth centuries, a significant proportion are older. Some were unplanned and simply grew spontaneously along fences; others are relics of long-vanished woodland. In certain places it is actually possible to date a hedgerow. As a general rule, one new tree or shrub species will colonize a

the north, and the birds are still disappearing at a rate of 10–40 per cent each year.

During the swinging sixties, intensive agriculture really got into its stride, offering much more for much less. Dairy cows became milk-producing machines, doubling their output to nearly 1300 gallons per cow per year. Livestock gained weight like sumo wrestlers, and our soils, freshly fertilized, produced ever more crops. In 1940 an acre of farmland yielded just over 5 tons of wheat, but today as much as 8 tons can be squeezed out of the same area.

As well as becoming more intensive, farms consolidated and grew larger in the postwar years. The number of farm holdings has halved since World War II, and just 3 per cent of all the farms in England and Wales now account for a third of the land. Farming had entered the big league. Specialization became the norm, with farmers tending to concentrate on a few key crops or livestock, but rarely both.

hedge every 100 years. A hedge with two colonists, therefore, is probably around 200 years old, while a hedge with 10 dates back to the Norman invasion. Although this formula doesn't hold true everywhere, it seems to work in Devon and the Midlands at least, and demonstrates just how long it can take to achieve a rich and varied hedgerow. Every time a hedge is grubbed out, a piece of living history is lost. Without fail, the older a hedge is, the more plant and animal species it contains. For many birds, small mammals and invertebrates, hedges are complete woodland habitats in miniature; some butterflies are almost entirely dependent on them, and many other animals use hedges as corridors between wilder patches of countryside.

Thankfully, many of our hedges survived the onslaught of intensive agriculture, and new hedges are now being planted. During the 1990s, nearly 25,000 miles (40,000 km) of hedgerow were planted or

(ABOVE) THE FARM LABOUR FORCE RAPIDLY DECLINED WITH MECHANIZATION AND THE USE OF CHEMICALS.

(OPPOSITE BELOW) THE BROWN HAIRSTREAK BUTTERFLY LAYS ITS EGGS ON YOUNG BLACKTHORN IN HEDGEROWS, AND THE CATERPILLARS FEED ON THE BLACKTHORN LEAVES.

(OPPOSITE ABOVE) OVER THE PAST 40 YEARS, MUCH OF UPLAND BRITAIN HAS BEEN DRAINED, PLOUGHED AND RESEEDED WITH GRASS TO SUPPORT INCREASING NUMBERS OF SHEEP. AS A RESULT, THERE HAS BEEN A RAPID DECLINE IN THE HEATHER AND OTHER DWARF SHRUBS THAT BLACK GROUSE NEED FOR FOOD AND SHELTER.

renovated – a distance 16 times longer than the Great Wall of China.

The postwar drive to increase productivity led to a surge in the use of synthetic fertilizers, herbicides and pesticides. From the 1950s onwards, farmers enthusiastically doused their fields with chemicals, and it's easy to understand why – 1 ton of chemical fertilizer has as much nitrogen as 25 smelly tons of manure and is considerably easier to apply. Herbicides made the removal of weeds by hoeing unnecessary, and raising crops became much less labour-intensive.

Fertilizers, herbicides and pesticides have, in one way or another, had a negative effect on wild animals and plants – either directly, by killing them, or indirectly, by removing their food supply. Of all the chemicals, one has an especially bad reputation: the insecticide DDT. Introduced in the 1940s, it was first used as a medical treatment for body lice, which caused an epidemic of typhus during and after World War II. Farmers found it equally useful as a weapon against insect pests, and it was lavished on fields. Other new insecticides, such as aldrin and dieldrin, were sprayed on seeds for further protection. All this chemical warfare certainly helped to improve yields, but the effects on wildlife were much further reaching than anyone anticipated. For birds in particular, it was nearly a disaster (*see box* 'Persistent Organic Pollutants').

But it's not all bad news when it comes to agriculture. During the last 20 years, the government has funded wildlife-friendly farming methods, and some 2 million acres (800,000 hectares) of farmland now benefit from environmental and conservation initiatives, such as the Countryside Stewardship Scheme. New organizations, such as the Farming and Wildlife Advisory Group (FWAG) and Linking Environment and Farming (LEAF), have been created, providing farmers with specific advice on making their land more wildlife-friendly.

Persistent Organic Pollutants

Insecticides such as DDT, aldrin and dieldrin are known as persistent organic pollutants (POPs) because they don't break down easily – they linger for years in the environment, causing untold damage. They have had a devastating impact on Britain's bird population and are all now banned.

POPs become incorporated into the food chain. When birds eat insects or seeds in a treated area, the chemicals enter their bodies and accumulate in body fat. Insect-eating fish also take in the chemicals, passing them on to fish-eating birds in turn. Because the chemicals accumulate in animals' bodies, they become increasingly concentrated and toxic higher up the food chain. At high levels they play havoc with hormones, disrupting reproductive processes. Many birds of prey and fish-eating birds have been affected by egg-shell thinning, which makes eggs too fragile to survive. British populations of sea eagles, peregrine falcons and sparrowhawks were badly hit by POPs in the 1960s and 70s, and seabirds, such as guillemots and gannets, have also suffered, as have cormorants and grey herons.

Since the chemicals were banned in the 1980s and 90s, bird populations have made a recovery. Peregrine falcons quickly bounced back to former levels, and the white-tailed sea eagle, which suffered from both persecution and pesticides, has been reintroduced to Scotland and now has a stable population. Sparrowhawks practically disappeared from eastern

England, where pesticide use was highest, but they had recovered completely by 1990, with 32,000 breeding pairs. Now you stand a good chance of seeing one swoop down onto your garden bird table.

(TOP) SPARROWHAWK
(ABOVE) WHITE-TAILED EAGLE
(LEFT) PEREGRINE FALCON

Coastal Squeeze

The Agricultural Revolution led to sweeping changes across the British countryside, but it wasn't just farmland that was transformed during the twentieth century. Some of the wildest parts of Britain have been tamed in the name of progress, including our long and rugged coast.

The UK has the longest coastline of any European country, and nowhere in England is more than 70 miles (113 km) from the sea. As an island nation, our coast is very important – both to us and to wildlife. In the north, rocky cliffs host colonies of screaming gannets, while the gentler shores of the south shelter grey seals in hidden sea caves.

Britain's coast is familiar and mysterious at the same time – a place of sandy beaches and sunny holidays, but also a place of windswept cliffs and inhospitable mud flats. Visit the Devon or Cornish coast on your holidays and you'll discover places of rugged spectacle and wonder; go to the east coast and you'll see vast, flat beaches that stretch into the distance. Surely these wild places could never be tamed or brought under our control? In fact, many coastal areas have undergone massive changes in the last 100 years, much of it with one intention in mind: to keep the sea out.

We like to think of our coast as a hard line on a map, something firm and definite, but this is far from the truth. Much of our coastline is 'soft', which means that the sea is slowly changing it, eating away at the land in some places and creating new land in others. Some of the most sensitive areas are around estuaries, where the low-lying land is vulnerable to floods (*see box* 'The Floods of 1953', p. 156). These dynamic landscapes are also the site of important wildlife habitats, but our efforts to keep the sea at bay can damage them.

After the catastrophic floods of 1953, sea walls were repaired and raised along England's east coast. Sea walls are not a new idea. They've existed in one form or another along much of our coast for at least 300 years, since farmers started building dykes to

(ABOVE) THE COASTS OF THE BRITISH ISLES ARE IMPORTANT TO GREY SEALS. THEY COME TO OUR SHORES IN AUTUMN TO JOIN BREEDING COLONIES ON INACCESSIBLE BEACHES AND SANDBANKS OR IN CAVES.

(OVERLEAF) ESTUARIES ARE IMPORTANT HABITATS FOR WADING BIRDS, WHICH SEARCH THROUGH THE SOFT MUD FOR SMALL WORMS AND OTHER INVERTEBRATES.

improve grazing on coastal marshes. Compared to the early defences, however, our modern sea walls are a marvel of engineering. Around 2000 miles (3200 km) of concrete, rubble and earth now protect eastern England from the North Sea. But how long can this hard engineering hold back the tide? Any rise in sea level increases pressure on the sea walls, and right now a couple of factors are doing just that. First, southeast England is sinking by about ⅛ in (2 mm) a year as the British Isles' tectonic plate continues to adjust to the end of the Ice Age. Second, global warming is raising the sea level, partly through thermal expansion of the water and partly through melting of inland glaciers. Sea levels are predicted to rise by between ⅛ in and ⅓ in (2 mm and 9 mm) a year, and this will put our concrete defences under great pressure. Our sea walls are already expensive to maintain, and many parts are crumbling. And these aren't the only problems – the walls also disrupt the balance of nature.

Sea walls contribute to the loss of an important habitat: salt marsh. This habitat occurs around most of the British coast, but the Thames Estuary, the Essex coast, Merseyside, Morecambe Bay and the Wash account for more than half of it. Salt marshes contain a rich variety of distinctive plants that share the ability to withstand occasional submersion in salt water. Such marshes tend to develop in sheltered, intertidal areas, such as estuaries in which the flow of water is gentle enough for mud to settle and stick. Plants gradually colonize the mud and stabilize it with their roots. A succession of different species grows between the low-tide level and dry land. Some, such as glasswort (*Salicornia*), thrive in the lowest parts of the marsh and can withstand being doused with salt water 600 times a year, whereas upper-marsh species, such as shrubby sea-blite, prefer only an occasional dip.

Salt-marsh plants provide food and microhabitats for a wide range of animals, including invertebrates, such as molluscs and worms. These, in turn, provide

The Floods of 1953

On the night of 31 January 1953, the forces of nature conspired to give Britain its worst floods in living memory. They claimed the lives of 307 people on England's east coast and killed more than 1000 in Holland. The event is still remembered as one of Britain's worst natural disasters.

The afternoon of Saturday 31 January was grey and cold, with strong, gusting winds, but there was no clue to what lay ahead. The North Sea coast was experiencing a high spring tide, and a deep depression was moving in from the north. The wind picked up, reaching gale force 10 and driving huge volumes of water across the sea surface. At the same time, the depression caused a 'storm surge' – a rise in the sea level caused by the drop in atmospheric pressure over the water. The deadly combination of gale-force wind, spring tides and storm surge raised sea levels more than 6 ft (1.8 m) above the normal high-water mark. Waves 20 ft (6 m) high crashed against the shore, breaching sea defences in 1200 places and inundating the land.

Canvey Island in Essex was the worst hit, with the whole island disappearing under water. Every house had to be evacuated. In all, 24,000 homes were damaged or destroyed, and more than 30,000 people were displaced. In Hunstanton, Norfolk, the pier, an amusement park and hundreds of caravans were swept away, and in Skegness the sand dunes disappeared. Power stations and hundreds of miles of roads and railways were put out of action, and thousands of acres of farmland were ruined.

(ABOVE) AFTER THE FLOODS OF 1953 MANY OF OUR SEA WALLS WERE REPAIRED AND BUILT UP EVEN HIGHER.

food for wading birds and waterfowl. The upper part of a salt marsh is often a high-tide refuge for birds that feed on the mud flats. Waders and gulls use it as a breeding site, and waterfowl use it as a feeding ground in autumn and winter. The Blackwater Estuary in Essex regularly has up to 30,000 waders, and more than 15,000 dark-bellied Brent geese fly all the way from western Siberia to spend winter here.

Essex has some of the best marshlands in Europe, but almost a quarter of them have been lost in the last 25 years. The problem is that the sea erodes the outer edge of the salt marsh, and with a hard sea defence behind it, the salt marsh can't re-create itself further inland. Scientists call it 'coastal squeeze'. The Environment Agency has come up with a radical solution. After spending years constructing sea walls, it has now started knocking holes in them. As well as helping to re-create salt marshes, the holes allow the sea to 'breathe', thus alleviating flooding in other areas.

Salt-marsh plants provide food and microhabitats for a wide range of animals.

(ABOVE) WHEN SEA LAVENDER BLOOMS IN SUMMER THE SALT MARSH BECOMES A PURPLE HAZE. THE COMBINATION OF DENSE PLANT LIFE AND RICH MUD PROVIDES PLENTY OF FOOD FOR A RANGE OF ANIMALS.

Road Rage

It all started in 1860 with the invention of the internal combustion engine. The ubiquitous motor car has changed our lives and our landscape, as it has done in the rest of the world. The network of cart tracks and cobbles that existed at the start of the twentieth century has become 250,000 miles (400,000 km) of tarmac, reaching into every corner of the British Isles.

Simply spotting a motor car was something of a novelty in 1900. Between the wars cars shot up in popularity, and today we have more than 30 million of them, in all colours, shapes and sizes. And we still haven't reached saturation. In the last 20 years, the number of cars in Britain rose by 60 per cent, with many households owning more than one. We simply can't get enough wheels.

So what effect has all this high-speed consumerism had on our countryside? For a start, our road network has had to expand to keep pace. After World War II, Britain's roads were neglected and inadequate. Road-building plans had been drawn up as early as 1938, but during the following years our civil engineers were tied up with building airfields and

The M40 Kestrels

A motorway engineer was carrying out routine checks on a bridge over the M40 when he noticed a kestrel using it as a nest site. It seemed like a curious place to raise chicks – just a few feet above speeding traffic – but in this fairly treeless part of Warwickshire, good nesting sites for kestrels are few and far between. This site certainly wasn't ideal. Kestrel chicks are very vulnerable when they make their first faltering attempts to fly at four weeks old, and the gusts of wind from lorries rushing past added to the dangers. Few of them survived.

But the M40's kestrel population was in luck. The Department of Transport and the Forestry Commission had been planting trees along the M40's verges for some time, only to find that the saplings were being killed by voles. The thick, unmown grass beside the motorway provided the perfect habitat for these small mammals, and there were hundreds of them scurrying through it. In fact, it was the voles that had attracted kestrels to the motorway in the first place.

The Department of Transport and the Forestry Commission hatched a plan. If they could encourage kestrels to breed, they could control the population of voles and so allow trees to grow on the verges. The kestrels clearly had a problem finding nest sites, so the Highways Agency tried mounting nest boxes behind the motorway's big blue road signs.

There are now about 30 nest boxes along the M40 in Warwickshire, some in trees and some behind road signs. The project is monitored by a bridge engineer who happens to be a fully licensed bird ringer. Many of the nest boxes are occupied every year, producing chicks with a good chance of successfully fledging. The project is such a success that there are plans to erect more boxes elsewhere on our motorway network.

So next time you're on the M40 near Banbury and you see a kestrel hovering high above, perhaps you should bear in mind that road engineers do a lot more than just laying down mile after mile of tarmac.

(ABOVE LEFT) DURING THE TWENTIETH CENTURY NEW CARS AND ROADS LEFT THEIR MARK. TODAY OUR ROADS TAKE UP AN AREA THE SIZE OF LEICESTERSHIRE.

other urgent military matters. Once peace was restored, the pressure was on to create bigger, faster roads linking the major towns.

The 1960s saw the largest road-building programme ever undertaken in Britain. In only 10 years, all the main roads were modernized and most of the motorways were born. Lancashire got there first when the 8-mile (13-km) Preston bypass was completed in 1958; it was the first stretch of motorway in Britain. The first section of the M1 was built the next year, and the motorway boom began. By 1972 we had 1000 miles (1600 km) of motorway, built at the average rate of 1 mile (1.6 km) per week, and many of the main traffic arteries we know today were in place.

The road-building blitz revolutionized life and opened new horizons for holidaymakers. Today's commuters might not feel there's much to celebrate when they're stuck on the M25, but back in the 1960s, day trips to previously unreachable destinations became perfectly possible thanks to motorways. It's ironic that these huge roads are now so clogged with traffic that journeys can take as long as they did in pre-motorway years, but the fact is we're now a highly mobile society. The average British adult travels more than 5000 miles (8000 km) a year by car – simply unthinkable in the first half of the twentieth century.

Roads have carved up the countryside into smaller and smaller parcels, and that isn't always good for wildlife. Many animals need large territories and room to travel, and roads get in their way. With around a million animals killed on British roads each year, there are also more obvious hazards. But cars and roads are a permanent fixture of our modern lives, and one to which we've all become accustomed. As always, there's a balance to be struck between development and conservation (*see box* 'The M40 Kestrels').

Conservation Heroes

It's clear that the twentieth century brought enormous changes to our landscape – that's nothing new. Throughout history, from the Stone Age to the Industrial Revolution, people have been changing our islands. But there's one thing that makes the twentieth century unique, and it's not so much the way we use our land as our attitude towards it.

A new eco-awareness emerged during the twentieth century, especially during the second half of it, as people grew uneasy about the increasingly intensive way our land is used. More people were taking day trips to the countryside to enjoy the great outdoors, and there was a growing concern that intensive agriculture, forestry, road-building and urbanization were not good for wildlife. Conservation bodies were set up, and membership of organizations such as Friends of the Earth rocketed. Today there are conservation groups throughout the country, many relying on help from volunteers.

Some campaigners became famous through their efforts. 'Swampy' was a firm favourite of the 1990s, an eco-protestor who campaigned against road and airport development by living in underground tunnels to hold up building work. Depending on your view, he was either a hero or a dropout. At the very least, he was noticed. Others prefer a more low-key approach, but one that is no less effective. These are the conservationists who quietly work away with little publicity, dedicated to making sure we don't squander our environmental inheritance.

On a clear, cold autumn morning on Kielderhead, you can see right across the border to Scotland. This part of Northumberland is a perfect example of upland Britain – rugged, quiet and beautiful. But like many other apparently wild places, it has not entirely escaped human interference. Historically, Kielderhead was wet and boggy, but over the last two centuries the bogs were drained and the heather was burnt for the benefit of grouse. Large parts of Kielderhead are now devoid of wet areas, and this seemingly small alteration has had a serious impact on the local wildlife.

The golden plover breeds in upland areas such as Kielderhead. It nests on open ground, and it feels safe only where it can see danger approaching from a mile off. There's another reason that plovers like these treeless moors: the boggy pools are full of the kind of aquatic invertebrates that plover chicks love to feed on. But recently the plovers haven't been seen in such great numbers, and the lack of pools seems to be to blame. The staff of Forest Enterprise and the local Northumberland Wildlife Trust decided to take action.

The obvious solution was to re-create the small pools that were once scattered across the hills, but this proved to be hard work. Mechanical diggers couldn't get here, so the workers faced a long climb and several back-breaking hours with a spade to dig each pool. One day, the engineers from Forest Enterprise were chatting about the problem and came up with a startlingly simple solution: dynamite. They were already adept at using explosives, and they were happy to turn their skills to conservation. What's more, it would be fun.

The engineers waited until autumn, when the birds had migrated to the coast for winter and it was safe to experiment. The first blasts were too powerful and the holes too deep for small chicks to wade in. After further experiments and adjustments, the engineers became expert at creating strings of shallow scrapes no more than a few feet deep. These soon filled with rain.

There are now quite a few pools at Kielderhead, each a miniature ecosystem complete with aquatic plants and invertebrates. It's a triumph of crude technology being used in a sensitive way, and it's one of hundreds of projects that are carried out around the country every year by our unsung heroes of conservation.

Closer to Home

There's a particularly desolate corner of Canvey Island in Essex that has been left untouched for more than 30 years. Lying in the shadow of an oil refinery, and littered with burnt-out cars, old bikes and rusting shopping trolleys, it doesn't look like anything to get excited about. But this unpromising 'brownfield site' holds a secret. It's something that's taken ecologists and developers by surprise, and recently there's been a buzz about the place.

It turns out that this site is home to at least 1400 wildlife species. It has more species per square foot than any nature reserve, and experts are calling it 'England's little rainforest'. The place is a haven for insects: it's home to five of our most threatened bumblebee species, 300 species of moth, the scarce emerald damselfly, and a weevil previously thought to be extinct. Badgers and water voles have made their home here, skylarks fly overhead, and in spring the place is a riot of orchids.

(ABOVE LEFT) THE GOLDEN PLOVER BREEDS ON UPLAND MOORS AND BOGS.

(OPPOSITE) CANVEY ISLAND: AN UNLIKELY WILDLIFE HAVEN.

(OVERLEAF) BLUETITS FEEDING ON A BIRDFEEDER. THIS IS AN EASY WAY TO HELP OUT BIRDS IN YOUR GARDEN, ESPECIALLY AT THE END OF WINTER WHEN FOOD IS SCARCE.

So what makes it so special? Interestingly, it seems that the unusual history of the site has contributed to its success as a top-notch wildlife habitat. In the 1960s the oil company Occidental planned to build a huge refinery here, and before construction began they dredged up tons of silt from the Thames Estuary and dumped 7 ft (2 m) of it all over what used to be a grazing marsh. This free-draining covering created the perfect soil for lots of wild flowers, and because the climate here is relatively dry, it also became a haven for insects. When the economy crashed in the 1960s, Occidental abandoned their plans, put a fence around the site, and left it alone. A few local lads used the area for trail biking and unwittingly added to its wildlife value. By constantly disturbing the ground, they kept some patches bare, which suits many insects. The place now has a mixture of different habitats: wetland, sandy areas, bare patches, short grass and scrubby bits. Ecologically speaking, it's very rich, but recently this brownfield site came under threat.

The Thames Estuary is a prime area for redevelopment, with thousands of new homes and businesses due to be built over the next 30 years, mainly on brownfield areas. The Canvey site looked ripe for renovation, and plans were submitted to turn it into a business park – a move that would undoubtedly jeopardize the rare mix of wildlife. Unusually, English Nature stepped in and brokered a unique deal. The developers have now agreed to build on only 20 acres (8 hectares) of the 600-acre (240-hectare) site, and their plans include many design innovations that will help the bugs. The remainder of the site will become a nature reserve, protected and managed for the benefit of both wildlife and the local community. But despite its new-found status, you won't find any 'keep off the grass' signs here. The kids who enjoy trail biking are likely to be encouraged to continue, since their activities have contributed to the diversity of the place. This is certainly conservation for the twenty-first century.

Brownfield sites are not the only urban places that provide a haven for wildlife. Some of our richest wildlife sites are even closer to home: our gardens.

There are around 15 million private gardens in Britain, covering about a million acres. Statistics suggest that three-quarters of the adult population has access to a garden, and that means that many of us have a fundamental role to play in protecting Britain's biodiversity. It's not just conservationists who can improve the balance of nature – every one of us can make a difference, even with a window box.

Scientists have recently come to the conclusion that gardens are very important to wildlife. Ecologists at the University of Sheffield carried out the first ever 'garden audit' in their city and counted some 175,000 green spaces. They discovered a wealth of tiny but important native animals, such as snails, spiders, beetles, wasps and crane flies, and an enormous diversity of trees, flowers, shrubs and grasses. Gardens are also full of small habitats, from ponds and nest boxes to trees and lawns. Think of each garden as a small patch on a larger quilt, and you begin to understand how, joined together, they constitute an important chunk of our 'countryside'.

One of the most surprising findings was that, whatever type of garden you have – manicured or messy – it still creates space for wildlife. But if you want to make sure your garden is a place where wildlife can really thrive, there are a few things you can do to give nature a helping hand (*see box* 'Creating a Wildlife Garden', pp. 164–5).

The last century has seen more changes in our landscape and way of life than any previous century. The two world wars affected the countryside every bit as much as they affected people, and with the advancement of technology and a growing population, the pressure on the land has never been greater. Juggling the demands of the population and the demands of wildlife has never been easy. At the moment, we are managing to keep our heads above water. But what of the future?

Creating a Wildlife Garden

No matter how small your garden, anything that increases the mosaic of habitats within it will benefit wildlife. If you build a pond, leave a few areas of uncut grass beneath hedges, plant a butterfly bush and stop using chemicals, you're well on your way. Here are some of the simplest things you can do to attract the birds and the bees.

Build a Pond

A pond will attract a wide range of new plants and animals to your garden. If you follow a few simple rules, you'll be watching dragonflies, damselflies, frogs, newts and birds in no time at all.

- Find a good sunny spot – ponds fare better when warm and unshaded.
- Don't place your pond too near to trees – the leaves will fall in and upset the oxygen balance.
- Make your pond in spring so that it can develop over summer and be established by autumn. Top it up in dry weather.
- When you dig the hole, include shallow 'shelves' for marginal plants.
- Line the hole with pool underlay and a good-quality pond liner.
- Aim for a good mixture of plants. White water-lily and frogbit, both of which have floating leaves, offer shade. Upright, or 'emergent', plants, such as flag iris, give dragonfly larvae a place to climb out. Place oxygen-producing aquatic plants under water to help keep it clear (use native species such as water milfoil).
- Always leave some parts of the pond's edge free of plants to allow access for animals.
- Avoid putting fish in a wildlife pond. Although a few small native fish will do no harm, exotics will eat everything from tadpoles to baby newts.
- To kick-start your pond life, get a friend to give you some adult plants and sludge from their pond. You'll acquire a good range of adult, larval and egg stages of aquatic invertebrates.

Go Organic

Using chemicals in your garden might get rid of one problem, but it can lead to others. Remember that insects are a vital part of the ecology of a garden, and if you want to attract larger animals, you'll need to keep the insects. Some insects may eat your plants, but they in turn are eaten by birds, mammals and each other. And plants depend on insects for pollination.

- Insects are very important in helping birds survive, especially chicks – a single brood of swifts eats about 20,000 insects per day. Avoid those sprays, and encourage your insects.
- If aphids are a problem, plant poppies or marigolds to attract hoverflies – one hoverfly larva can eat about 600 aphids during its development.
- Use organic fertilizers, which feed soil bacteria and keep your ground healthy.

Feed the Birds

There are many ways to feed birds in your garden, from letting insects flourish to buying a bird table. Remember that birds are most in need of help in late winter, when wild food is hard to come by.

- Grow fruit trees – the leftover fruit and windfalls provide food for many birds.
- If you grow fruit bushes, such as raspberries, don't cover them entirely with netting. Leave some of the fruit exposed for the birds to eat. Yes, I know it goes against the grain, but what would the gardener's life be like without the song of the blackbird of an evening?
- You can grow specific plants for seed-eating birds. Goldfinches love thistledown, and sunflowers are good for greenfinches. You may need to support the heads of sunflowers until the seeds ripen.
- Put out bird food regularly. Use a good mixture of seeds, fruits, fatty foods, peanuts and wholemeal breadcrumbs. RSPB-recommended bird foods are very good.
- Make sure birds always have fresh water, too.
- Place a bird table near a window so that you can enjoy watching your visitors.
- Avoid placing a bird table near bushes where cats can hide – birds are vulnerable when approaching the table.

Attract Those Butterflies

Butterflies are likely to find their way into your garden come what may, but if they like what they find, they'll stay.

- Like hoverflies, honeybees and bumblebees, butterflies need sugar-rich nectar. Plant a 'nectar trap' full of scented, nectar-rich flowers (single rather than double) in a sunny corner of your garden. Nectar-rich species include buddleja, *Sedum spectabile* (not 'Autumn Joy'), *Verbena bonariensis*, red valerian, *Echinacea purpurea*, lavender and sweet rocket.
- Different species of butterfly emerge from hibernation at different times of year, so to benefit as many as possible, you should phase your planting of butterfly-friendly flowers.
- Grow plants that caterpillars can feed on. Butterflies lay their eggs on many different plants, but four of the prettiest butterflies like stinging nettles. If you let a patch of stinging nettles thrive in a sunny and sheltered spot, you'll be providing caterpillar food for commas, small tortoiseshells, peacocks and red admirals. Nettles can spread, so keep them well away from your flower borders.

(OPPOSITE) GARDEN PONDS ARE A WILDLIFE HAVEN. (ABOVE) PAINTED LADY BUTTERFLY (LEFT) BEES ARE IMPORTANT POLLINATORS. YOU CAN ATTRACT THEM BY PLANTING A NECTAR-RICH FLOWER BORDER IN A SUNNY SPOT.

The Future

THE HARDEST THING FOR HUMANS TO CONTEMPLATE IS TIME. HOURS, DAYS, WEEKS AND YEARS ARE WITHIN OUR COMPASS, BUT THOUSANDS AND MILLIONS OF YEARS ARE NOT. PERHAPS THAT'S WHY PREHISTORY SOMETIMES LEAVES US COLD. IT'S NOT A LACK OF INTEREST – IT'S AN INABILITY TO COMPREHEND. BRITAIN'S OLDEST ROCKS ON THE ISLE OF LEWIS ARE NEARLY 3 BILLION YEARS OLD. HOW CAN ONE GET TO GRIPS WITH THAT? IT'S JUST NOT POSSIBLE. BUT BY TRYING TO UNDERSTAND WHAT THOSE ROCKS HAVE

WITNESSED, WE CAN AT LEAST HAVE SOME APPRECIATION OF OUR ISLANDS' ANTIQUITY. SINCE THEIR BIRTH, THEY HAVE SURVIVED MANY TRANSFORMATIONS OF THE LAND – GREAT MOUNTAIN RANGES HAVE RISEN AND CRUMBLED BACK TO THE GROUND, VOLCANOES HAVE ERUPTED, AND THE WHOLE OF THE BRITISH ISLES HAS BEEN SMOTHERED BY OCEANS, TROPICAL FORESTS, SANDY DESERTS AND ICE. THE ROCKS OF LEWIS HAVE SEEN IT ALL.

As we go about our daily life, whether we are at work, at home or in the garden, it's easy to take our history for granted, or simply to ignore it as being a bit scary. After all, in our lifetime, we simply won't see mountain ranges being thrust up from the continental shelf, and neither will we feel our islands moving slowly across the globe. For all practical purposes, this is where we are now, and this is where we will stay. But the British Isles are changing and always will change. The trick is to be fascinated by it rather than terrified.

So what are the changes going to be? What does the future hold for our islands? The simple answer is that no one really knows. Scientists can and do anticipate a whole host of disasters that could strike us at some point. Tidal waves half a mile tall might crash into our coast; meteorites bigger than the one that killed the dinosaurs might plough into us; and 'supervolcanoes' could explode with enough force to obliterate whole continents. But then again, they might not. With all of these predictions, it is far more likely that none will happen than that any ever will.

Although we can't rule it out, it seems most unlikely that any of us will live to see the British Isles changed by an apocalyptic disaster. But there are more subtle changes that we can witness, changes that are going on right now. And the evidence is all around us.

Changing Seasons

> I wandered lonely as a cloud
> That floats on high o'er vales and hills,
> When all at once I saw a crowd,
> A host, of golden daffodils;
> Beside the lake, beneath the trees,
> Fluttering and dancing in the breeze.
>
> *William Wordsworth (1770–1850)*

William Wordsworth is one of Britain's most famous poets, and this is surely his most famous verse. It conjures up images of an idyllic spring morning spent strolling beside Ullswater, but thanks to William's sister Dorothy, we know that the day was not quite so serene. Dorothy Wordsworth was herself a keen writer, not of poetry but of her own private journal. Her recollections of that day are a little different from William's: 'the wind seized our breath, the Lake was rough'. But the important thing about her journal entry is not her lack of poetic licence; it's the date. She and William had encountered the now-famous daffodils on 15 April 1802.

In recent years the Wordsworth Trust has held a daffodil fair in honour of the poet, and each year it gets a little earlier. The fair now takes place in the middle of March, so daffodils are flowering a whole month earlier than 200 years ago, when Wordsworth wrote his immortal botanical poem.

(ABOVE) DAFFODILS STILL FLOWER ON THE SHORE OF ULLSWATER IN THE LAKE DISTRICT BUT NOW THEY APPEAR A MONTH EARLIER THAN IN WORDSWORTH'S DAY.

(OPPOSITE) ON 16 MARCH 2880, ASTEROID 1950DA, MEASURING 0.6 MILES (1 KM) WIDE AND WEIGHING OVER A BILLION TONS, MIGHT CRASH INTO EARTH. BUT IT PROBABLY WON'T: THE ODDS OF IT HITTING ARE 300-1 AGAINST.

(PREVIOUS PAGE) ATMOSPHERIC POLLUTION WILL AFFECT OUR FUTURE CLIMATE.

(ABOVE LEFT) WHITE DEAD-NETTLE (*LAMIUM ALBUM*) NOW FLOWERS EIGHT WEEKS EARLIER THAN IT DID 50 YEARS AGO.

The daffodils are not alone. Thanks to diarists such as Dorothy Wordsworth, naturalists such as Gilbert White of Selborne, and a great many other enthusiasts, there are detailed records of flowering dates going back more than 200 years. Close examination of records for 385 wild-flower species has revealed a clear trend. On average, spring flowers appeared 4½ days earlier in the 1990s than in the period between 1954 and 1990, and as many as 200 species now flower two weeks before they did 50 years ago. Some plants are flowering even earlier: the opium poppy appears 20 days sooner than in the 1950s, the lesser celandine 21 days, the greater stitchwort 25 days, and the ivy-leaved toadflax 35 days. But the winner is the white dead-nettle, which is so common on road-sides: it flowers 55 days before it did 50 years ago – so early, in fact, that it's now flowering in winter. And it's not just flowers that are appearing before they used to. Oak leaves are sprouting 10 days earlier, and the leaves of hornbeams, horse chestnuts and sycamores are all opening sooner too.

Animals are also racing to emerge from the winter doldrums. Frogspawn is appearing earlier, migrant birds are arriving ahead of schedule, and hedgehogs are awakening sooner from their winter dormancy. Everything points to the fact that spring is happening earlier. And if spring is changing, what about the rest of the year?

> Remember, remember the fifth of November
> Gunpowder treason and plot.
> I see no reason why gunpowder treason
> Should ever be forgot.
>
> *Anon*

Fireworks have certainly changed in the last 50 years. They've got bigger, better and, best of all, safer. Gone are the lethal jumping crackers that used to make us leap into the air. No longer do you light the blue touchpaper – now there's a fuse. But there's something else that's different about the fifth of November: the bonfires burn more fiercely. Why? It's all down to the material. You can build the best structure possible with logs and planks and pallets, but if you go and dump a heap of leaves on it, your fire will make more smoke than flames. These days, bonfires are better because there are fewer leaves – most are still on the trees. Today's trees hang onto their leaves well into November. Gardeners may also have noticed that grass is more active than it used to be in late autumn. Some of us even need to mow our lawns all year around.

Autumn is happening later and spring is arriving earlier – all of which points to the fact that the British Isles are warming up. And as newspaper articles seem to keep telling us, every summer is the hottest on record. In 2003 Britain recorded its hottest day ever, with temperatures peaking at 38.5°C (101.3°F) in Faversham in Kent. This was not a one-off – the three warmest years in the past 600 have occurred since 1990. Globally speaking, the 1990s were the warmest decade in the last century. And the last century was the warmest in the last millennium.

There seems to be little doubt that the British Isles and the rest of the world are heating up. Global warming is happening, but then, as those tropical interludes in the Ice Age prove, it's happened before. It's just that now the activity of people is a major contributory factor.

The Future Is Warm

In 2001 some 39 million Britons spent their holidays abroad. The most popular destination was Spain, which accounted for almost 30 per cent of foreign holidays. Close behind was France, which just pipped Greece in the popularity stakes. People go abroad for all sorts of reasons, but the main thing we want is sun – and all the things that come with it: warm seas, siestas on the veranda, sun-ripened tomatoes straight from the vine, fresh fruit and local wines. For many of us, holidays in warm climes are as close to paradise as you can get.

So if global warming means we can snorkel off Blackpool without a wetsuit and return home to a British chardonnay, shouldn't we be celebrating? After all, we've had hot spells in the past and nothing bad happened then. Around 1000 years ago, the world went through what scientists call the 'Medieval Warm Period'. For Britain it was a time of long, hot summers and short winters. The population more than tripled in 200 years because of bumper harvests, and people weren't just harvesting the usual staples – the Domesday Book has records of more than 80 vineyards across England. It all sounds rather pleasant really. If it happened again, we could grow olives and harvest our own tea, and perhaps the England cricket team could practise without rain and win the Ashes back ... no, let's not get carried away!

This time it looks as though temperatures are going to be higher than during the medieval hotspot. In fact, they could be unbearably high. Some climate-change experts think summer temperatures could exceed 40°C (104°F) on a regular basis. Such temperatures would have extraordinary repercussions for our wildlife.

For some species, warming is an opportunity; for others it's a death sentence. Species that like the heat are currently restricted to southern England, but as the British Isles warm up, they will spread north. It's been estimated that every 1°C (1.8°F) rise in temperature means an animal can, or has to, move about 100 miles (160 km) north or 300 ft (90 m) uphill. And as the south warms up, animals and plants that once found the whole of our islands too cool will be able to move in from Europe.

For cool-climate specialists, especially those that are already at the southern limit of their distribution, higher temperatures will be disastrous. Some might survive by moving north or to higher ground, but once they reach the hilltops or the coast, there will be nowhere left to go; global warming will mean an end to their residency in the British Isles. Those that manage to survive on mountains will face new problems, such as dwindling numbers and fragmentation of the population into small groups that are isolated by inhospitable valleys.

The comma butterfly is one of the species that

Why Is the Climate Changing?

Global warming seems to have a number of causes, both natural and unnatural. The Earth's climate has often changed in the past, and the British Isles have been through tropical periods, desert periods, and periods when they were covered by ice. Even in the Ice Age, the climate swung back and forth between freezing and pleasantly warm. But experts are worried that the climate is now changing faster than ever before, and this suggests that human interference is to blame.

The main culprit is the greenhouse effect. Just as the glass panes in a greenhouse trap heat reflected off the sun-warmed ground, certain gases in the Earth's atmosphere trap heat rays that would normally escape into space. Such gases include water vapour, carbon dioxide, nitrous oxide and methane. The levels of these 'greenhouse gases' – especially carbon dioxide – is rising, largely because we burn so much fossil fuel to power our modern lifestyles. Ever since Abraham Darby sparked off the Industrial Revolution, the amount of greenhouse gas pumped into the atmosphere has increased phenomenally.

Nature and man have combined to increase temperatures globally. How far this progression continues depends on how much more greenhouse gas is pumped into the air. This, in turn, depends on all sorts of factors, including global population, alternative energy sources and changes to the economy. Because there is so much uncertainty, the UK Climate Impacts Programme has come up with varying scenarios based on different levels of emission. It makes for alarming reading. Based on the high-emissions scenario, by 2080 summer temperatures in Britain could be 5°C (9°F) higher than today. The consequences of such a change to our climate would be enormous.

(ABOVE) POLLUTION HAS INCREASED STEADILY IN THE LAST 200 YEARS, AND EXPONENTIALLY IN THE LAST 50.

(LEFT) PREDICTED INCREASE IN SUMMER TEMPERATURES DURING THE 2020s, 2050s AND 2080s FOR THE LOW-EMISSIONS AND HIGH-EMISSIONS SCENARIOS.

seem to be benefiting from a rise in temperature. Its range has increased by more than 60 per cent in the past two decades, and it looks likely to return to Scotland after an absence of more than a century. Most of Scotland and Ireland will probably also gain the speckled wood butterfly and the large skipper by the middle of the century.

By contrast, some northern butterfly species are declining dramatically as they disappear from southern parts of their range. The outlook is also bleak for upland species, such as the large heath and northern brown argus. The mountain ringlet butterfly, which is found only in the Lake District and western Scotland, will probably find no suitable habitats in the British Isles by 2050.

It's a similar story with birds. Once again, the northern and upland species face an increasingly uncertain future. Birds such as the ptarmigan, snow bunting and dotterel could disappear from the Cairngorms and become extinct in Britain. Already at the southern extremity of their ranges, their only option will be to leave the country for good.

But it's not all doom and gloom. Birds are highly mobile and can quickly adapt to changing conditions. As we lose some species, others will move in. It's already happening. The little egret is now frequently

(ABOVE) THE LARGE SKIPPER WILL CONTINUE TO EXPAND ITS RANGE AS BRITAIN KEEPS WARMING UP.

(LEFT) A GREAT WHITE EGRET, PHOTOGRAPHED IN CHESHIRE – FURTHER NORTH THAN HAD PREVIOUSLY BEEN RECORDED.

(ABOVE) CALEDONIAN PINE FOREST WITH SCOTS PINES (*PINUS SYLVESTRIS*) AT ABERNETHY FOREST, SPEYSIDE. SUCH FORESTS ARE VULNERABLE TO CLIMATE CHANGE.

(OVERLEAF) BEECH WOODLAND COULD DISAPPEAR IF GLOBAL WARMING MAKES BRITAIN'S SUMMERS HOT AND DRY.

Caledonian pines once cloaked much of Britain, but as our climate warmed, they retreated north to the colder, higher glens.

seen in England since its arrival from southern Europe some 20 years ago. Bee-eaters have recently arrived from France, as has the cetti's warbler, which now breeds in our southern wetlands. In the same way, resident birds, such as the nuthatch, serin, yellow wagtail, reed warbler, lesser black-backed gull and Dartford warbler, have all started to expand their ranges, moving north with the heat.

While most animals at least have the option to migrate quickly to find more agreeable conditions, plants have no such luxury. One of the most vulnerable is the Caledonian pine, a cool-climate species that was one of the first trees to arrive after the Ice Age. Caledonian pines once cloaked much of Britain, but as our climate warmed, they retreated north to the colder, higher glens. Today, places such as Glen Affric and the valleys of the Cairngorms are their last stronghold in Britain, but even here they are under threat from rising temperatures.

Surprisingly, our glorious beech woods are also threatened by global warming – not by higher temperatures per se, but by the dry weather that warming might bring. Beech trees have shallow roots, an adaptation for life on the thin, chalky soils of southern England. When the water table drops in hot summers, the roots can no longer get enough water. It took years for England's beech trees to recover from the great drought of 1976, and many were blown down in the storm of 1987. A succession of droughts could spell the end for one of our most magnificent tree species.

Once again, a problem for some species is an opportunity for others. Tree species from southern Europe would be ideally suited to a warmer Britain. After the Ice Age ended, trees spread into our country across the land that once joined England and France; today they'd need a helping hand across the English Channel. Importing Continental trees might be the best way to maintain the health of our woodlands. French and Spanish oaks could supplement our native oak, which copes less well with hot, dry weather. Species such as walnut and almond would thrive here, and the sycamore – a medieval introduction from southern and central Europe – would become far more widespread.

The idea of a warmer Britain seems very alluring – think of those glorious summers and warm seas, sitting in outdoor cafés and driving convertibles. But the reality is not so perfect. A warmer world may also be a sicker world, for all kinds of reasons.

For a start, there would be more deaths from heatstroke, particularly in cities. An estimated 13,600 people died in France in August 2003 when a heatwave brought the hottest weather for 60 years and the mercury hit a sweltering 40°C (104°F). The heatwave also hit Britain, causing 907 more deaths than usual during one week in mid-August – possibly a taste of things to come. Sunnier weather would also mean thousands more deaths from skin cancer and thousands more cases of cataracts.

Insect-borne diseases could spread from hot countries to Britain, giving us dengue fever and yellow fever. Experts reckon that a mild strain of malaria could become established in parts of the British Isles for up to four months of the year. Hot weather also increases the incidence of gastrointestinal bugs, including the bacteria that cause typhoid and cholera, with clinical consequences ranging from mild stomach upset to fatal gastroenteritis and septicaemia.

Unduly pessimistic? No. The challenge for society is to adapt to the new threats rather than simply lying down and accepting them. They will become the new millennium's equivalent of diphtheria and tuberculosis. We are lucky that modern medicine at least has a chance of keeping pace with them – an advantage not possessed by our forebears.

The New Invaders

A few Spanish oaks and walnut trees might make a pleasant addition to the British countryside, but not all new arrivals would be so welcome. Foreign pests are already threatening to extend their range into the British Isles and infest our local flora. One of the most feared is the Asian longhorn beetle – a devastating pest of many hardwood trees. It's a voracious little blighter. It spends most of its two-year life as an inch-long larva buried deep in the tree, slowly eating its way through the wood. The constant munching ultimately kills the tree. Worryingly, the beetle is a fast breeder – adult females lay up to 60 eggs in almost as many trees.

Increased trade with China, where the insect originates, raises the chance of it entering Britain. It's already reached the USA and killed thousands of trees there. If it escaped here, it could attack horse chestnuts, poplars, willows and fruit trees, not just in the wild but in our streets and gardens. Experts think it could survive in many parts of the country; if it became widespread, it could prove impossible to eradicate. Forestry Commission inspectors are stationed at all major UK ports to seek out and destroy the so-called ALB. So far, inspectors have recorded more than 50 cases of imported material showing signs of infestation. It may be only a matter of time before the first outbreak is recorded.

Invading species are nothing new. As we saw in earlier chapters, much of our islands' wildlife is not

strictly native. Poppies, horse chestnuts, sycamores, mandarin ducks, pheasants, garden snails and rabbits were all introduced by people, long after the English Channel formed and isolated Britain's true natives from Europe. In fact, of Britain's 49 wild mammal species, about 21 are introduced, including eight of our largest: the wild goat, fallow deer, Sika deer, Indian muntjac, Chinese muntjac, Chinese water deer, Bennett's wallaby and the reindeer. These mammals originate from as far afield as India, Australia, China and the Arctic, which shows just how cosmopolitan the British Isles have become.

As global warming makes our shores increasingly hospitable to foreign plants and animals, we can expect an increase in new arrivals. Like many of our long-established immigrants, the new invaders may prove to be mostly benign. However, a few could wreak havoc on the countryside, as experience has taught us. The problem with introduced species is that the diseases, predators or competitors that would normally keep them in check are sometimes absent. As a result, they can multiply without restraint and make hay at the expense of our native animals and plants. A few examples demonstrate how devastating introduced aliens can be.

The American mink was first brought to Britain in the 1930s to stock fur farms. Some inevitably escaped, and in 1957 they bred for the first time in the wild, in Devon. Since then, they have spread across the country, filling a niche in the ecosystem for an aquatic carnivore. They have had a devastating effect on many of our native animals. More than 90 per cent of Britain's water voles have disappeared because of mink. Mink even spread to the Outer Hebrides, where it has decimated populations of ground-nesting seabirds. When mink finds a gull or tern colony, it's unusual for any of the chicks to fledge.

Dutch elm disease arrived here accidentally in the early 1970s in a shipment of infected timber from Canada. The disease is caused by the fungus *Ophiostoma ulmi*, which is spread by a particular species of bark beetle. By 1973 the epidemic was killing elms throughout the country, obliterating one of our most distinctive species. More than 17 million trees were killed in just a few years. Thirty years later, the magnificent specimen trees are long gone, but the

(ABOVE LEFT) THE BEAUTIFUL YET DEVASTATING ASIAN LONGHORN BEETLE.

Reintroducing Lost Species

The British Isles have lost many species over the centuries, mainly because of man's intervention. Many of our large mammals were hunted to extinction, either for food or because they preyed upon livestock or interfered with managed habitats. Bears, wolves, wild boars, beavers and various birds of prey, including the great bustard and the sea eagle, all disappeared in the last few centuries. Some of these species have been reintroduced in an effort to re-create a more complete native fauna. The red kite and the sea eagle have been very successful, and the wild boar is back in the form of feral pigs, which are thriving in parts of Kent and Sussex.

But other reintroductions are controversial. The European beaver disappeared from Britain about 400 years ago after being hunted for fur, meat and its scent glands, which were thought to cure headaches. Now conservationists want to reintroduce it by releasing beavers from Norway into Scottish rivers and the wetlands of East Anglia. However, the proposals have met with resistance from farmers and landowners, who are anxious about the effects beavers might have on drainage systems and salmon stocks. For the moment, the proposal is on hold.

Even more controversial is the plan to reintroduce wolves to Scotland. In theory, wolves could make a comeback. The Scottish Highlands cover some 10,000 square miles (25,000 sq km), and the area is one of the most thinly populated in Europe, so there is ample space for wolves to establish territories and form packs. And the region is well stocked with wild prey; in fact, the local population of red deer has exploded in the absence of natural predators. But there are also sheep farms scattered throughout the Highlands, and there would inevitably be conflict with farmers if wolves preyed on their livestock.

Other countries have got round this problem by setting up compensation schemes to pay farmers for lost livestock. Perhaps a similar scheme in Scotland would allay farmers' fears and pave the way for the reintroduction of this impressive predator.

(TOP) WILL WE HEAR WOLVES HOWLING ON THESE SHORES AGAIN ONE DAY?

(ABOVE) WILD BOAR ARE NOW RELATIVELY COMMON IN PARTS OF ENGLAND.

(LEFT) EUROPEAN BEAVERS MAY SOON BE FOUND THROUGHOUT BRITAIN ONCE AGAIN.

elm still hangs on in hedgerows, now rarely reaching more than 15 ft (4.5 m) tall before succumbing to a new attack of the fungus.

Japanese knotweed was brought from Asia in the mid-nineteenth century to be grown as an ornamental and fodder plant. Today people call it Britain's most dangerous plant. It's incredibly invasive and aggressive. Its bamboo-like shoots can grow 4 in (10 cm) a day and reach a height of 10 ft (3 m), forming an almost impenetrable thicket that suffocates and smothers all other ground plants. Once established, knotweed is very difficult to get rid of because it can regenerate from fragments of underground stem as small as thumbnails. Surprisingly, all the plants in Britain are genetically identical females. In fact, the whole population is made up of clones of one individual. Identical clones are also found in continental Europe and North America, so Japanese knotweed is arguably the biggest female organism in the world. A form that is able to produce seeds has recently been discovered, but even this cannot keep pace with the plant's rate of vegetative reproduction. It comes as something of a relief to gardeners.

Wet Winters

Summer droughts and soaring temperatures could make life in the cricket season pretty unbearable – for us and for our flora and fauna. But global warming is not going to be one long, hot summer. Winters will be affected too, and not as you might expect. Certainly, they'll warm up, which is good news for gardeners and bad news for skiers. But that's not the only change. Whereas summer rain is going to become scarce, winter rain is likely to increase. The forecast for future winters is rain, rain and more rain.

Climate models predict that by 2080 parts of the British Isles can expect at least 30 per cent more winter rain than they currently receive. It's a trend

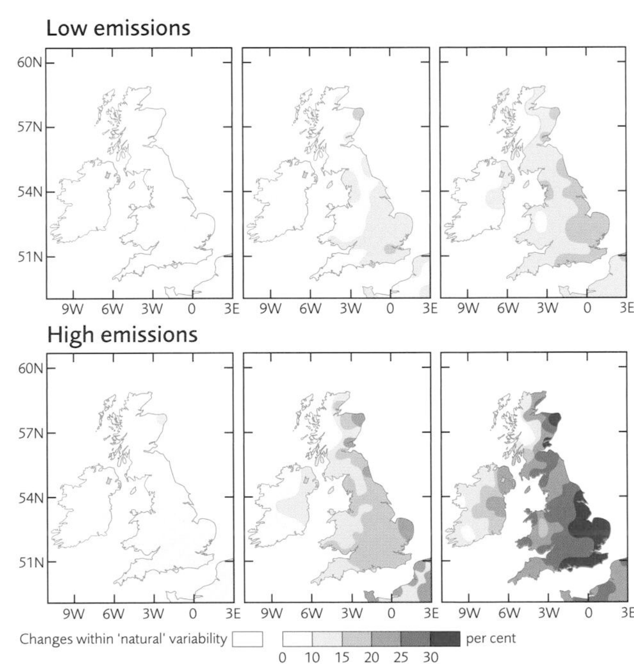

we've already noticed. The year 2002 was one of the warmest in history, but it was also one of the wettest; the same was true of 1998. Winter floods now seem to be a regular occurrence in much of Britain. Most of our rivers (including the Ouse, the Severn and the Trent) have broken their high-water records in this millennium and they will probably break them again. Nature reserves, parks and farmland will all be threatened by an increase in flooding – as much as 61 per cent of the best farmland in England and Wales is predicted to be affected. Who knows what's going to happen to house prices, let alone insurance premiums!

Climatologists tell us that we can also expect more 'extreme' weather. This means that those torrential downpours, stifling heatwaves and violent storms – all the things we think of as freak events – are going to be happen more often.

As the global temperature rises, so too will the

By 2080 parts of the British Isles can expect at least 30 per cent more winter rain than they currently receive.

(ABOVE) OCTOBER 2001 BROUGHT HEAVY FLOODING TO PARTS OF THE BRITISH ISLES.

(OPPOSITE ABOVE) AN INFESTATION OF JAPANESE KNOTWEED GROWING IN A SWANSEA CEMETERY.

(OPPOSITE BELOW) PREDICTED INCREASE IN WINTER RAINFALL DURING THE 2020s, 2050s AND 2080s FOR THE LOW-EMISSIONS AND HIGH-EMISSIONS SCENARIOS.

sea. Warm weather is melting the world's ice – more than 85 per cent of the planet's glaciers are already retreating. There's no better example of this than the Blomstrandbreen glacier on Svalbard, a remote Arctic island 375 miles (600 km) north of Norway. Since 1928 this glacier has retreated more than 1¼ miles (2 km). All over the world, ice is melting, and water that had been locked up on land in glaciers or ice sheets is flowing into the sea. Sea levels are rising for a second reason. As the global climate warms up, the oceans will expand in volume because sea water, like most things, expands when it heats up.

How the rising sea will affect our islands isn't clear. The outcome depends not only on how quickly the world warms up, but also on what happens to our land. Some parts of Britain are rising and others are sinking. This slow change in land level is a legacy of the Ice Age. Most of Scotland and northern England is still rising – the land is rebounding now that the massive weight of ice has gone. But the south is sinking. In essence, Britain is tilting. As a result, the consequences of sea-level changes will vary throughout the country.

Southeast England will be worst hit because it's sinking fastest. Taking into account both the rising sea and falling land, the effective sea level could rise by nearly 3 ft (90 cm) here by 2080. Housing, industry, tourism and wildlife would all be affected. At least 62 Sites of Special Scientific Interest on the English and Welsh coasts are threatened, including some of the largest wildlife areas in Britain. Salt marshes and mud flats – habitats that help to protect the coast from erosion – are most at risk. It sounds drastic, but luckily time is on our side. The seas are rising so slowly that we have more than enough time to prepare ourselves and protect the coast. People are already raising sea defences and setting aside areas of land that can be safely flooded. With time, effort, care and capital, we can prepare for the rising seas and minimize environmental damage.

(ABOVE) THESE TWO PHOTOS WERE TAKEN 84 YEARS APART IN EXACTLY THE SAME LOCATION IN SVALBARD. NOTICE THE EXTENT TO WHICH THE ICE HAS RETREATED.

What we can't prepare for are the unexpected storms. The sea will be most dangerous when high tides combine with violent storms, as happened during the great floods of 1953, when 307 people lost their lives. Floods of this scale currently happen only once every 100 years or so, but in the future they might happen every few years.

We obviously need to know by exactly how much sea levels will rise, and by when. The worst-case scenario prediction of a 3-ft (90-cm) rise in southeast England by 2080 assumes that nearly all the world's glaciers will melt, but not the polar ice caps. The reasoning is that polar ice caps are so huge that they create their own cold buffer zone, insulating them from the effects of global warming. But not everyone agrees with this idea. According to some scientists, the west Antarctic ice sheet – a slab of ice eight times larger than Britain and more than a mile thick – is unstable. And it has melted before. About 14,000 years

Warm weather is melting the world's ice – more than 85 per cent of the planet's glaciers are already retreating.

(ABOVE) STORMS WILL HIT OUR COASTS MORE FREQUENTLY IN THE FUTURE AND MAY WELL BREACH OUR IMPROVED FLOOD DEFENCES.

ago, as the Ice Age was ending, parts of the west Antarctic ice sheet seem to have collapsed and raised sea levels suddenly. At present, the massive east Antarctic ice sheet is definitely staying put, but if the west sheet collapsed, sea levels could shoot up 20 ft (6 m), which would be catastrophic for the British Isles. Thousands of square miles of land would disappear under the sea, and our coastline would change completely.

But there's no need to panic just yet. Indeed, no one is absolutely certain whether this will happen at all, but then awareness is everything. I don't want to appear dismissive, but even if it does happen, it is certainly not going to occur in our lifetime.

The Future Is Cold

Having convinced you that it's time to install air conditioning, sell the skis and buy a book on wine-making, I should point out that global warming could have precisely the opposite effect on the British Isles. It might make us freeze.

As the world warms up, melting Arctic ice and glaciers will add fresh water to the northern Atlantic, and some scientists think this could shut down the Gulf Stream. The Gulf Stream is the warm ocean current that keeps the British Isles about 10°C (18°F) warmer than we deserve to be, given how far north we are. It is part of a much larger ocean current system that circulates water through the Atlantic, taking it from the tropics to the poles and back again. When the Gulf Stream reaches the far north (where it's called the North Atlantic Drift), the salty water gets cold, which makes it dense and heavy. It sinks to the bottom of the ocean and turns back south towards the equator. If too much fresh water enters the north Atlantic from melting ice, the water won't be salty and heavy enough to sink. As a result, the whole circulation system could change and the Gulf Stream might move south or shut down altogether.

An end to the Gulf Stream would be bad news for our islands. Temperatures would plummet. Winters would be an estimated 11°C (20°F) colder, with blizzards a daily occurrence. The soil would freeze solid for much of the year, and icebergs would drift around our coast. Ports might freeze over, and we'd have to grow food in giant greenhouses if importing it became impossible. It might seem far-fetched, but it wouldn't be the first time since the Ice Age that the British Isles have been struck by a long bout of Arctic weather (*see box* 'The Little Ice Age').

If it happened again, we'd have to look on the bright side: we'd probably become a top winter-sports location. The Cairngorms, the Pennines, Snowdonia and the Chilterns – in fact, all our mountains and hills – would be ideal for skiers and snowboarders. We might even get to host the winter Olympics!

A chilly climate would lead to big changes in the countryside. The fragments of Caledonian pine forest in Scotland would expand and spread across Britain, and the call of the capercaillie would reverberate as never before. Our native amphibians and reptiles would disappear, migrant birds would avoid our shores, and much of our woodland would succumb to the cold; in the Little Ice Age, many trees actually split apart because the cold was so intense. Mountaintops might be cold enough for snow to stay all year round. Slowly this snow would build up and turn into ice, forming glaciers. Once again, ice would carve its way across the land – though in all likelihood, this would probably happen only on the tallest, north-facing slopes in northern Scotland.

Ice can be self-sustaining. It reflects the sun's warmth back into space, creating colder weather (provided there's enough ice). Perhaps this ability could trigger the onset of a new ice age. After all, glacial periods have come and gone for the last couple of million years, and there's no reason to think that this cyclic process has stopped. According to some experts, we're technically still in the Ice Age, albeit in one of the warmer 'interglacial' periods.

The Little Ice Age

(LEFT) FROST FAIRS WERE ALL THE RAGE WHEN THE THAMES FROZE OVER. THERE WERE STALLS SELLING CAKES AND BRANDY, PRINTING PRESSES WERE WHEELED ONTO THE ICE, AND OF COURSE THERE WERE GREAT SKATING RACES.

(BELOW) TEMPERATURE CHANGES IN THE BRITISH ISLES OVER THE PAST 1050 YEARS. NOTICE THE MEDIEVAL WARM PERIOD AND THE LITTLE ICE AGE.

Imagine a noisy, bustling market. There are stalls selling roast ox, mulled wine, gingerbread, brandy balls, black pudding and pancakes. There's a printing press squeezed between a puppet show and a travelling theatre. Nearby, people are placing bets on a donkey race, and a crier is announcing that a fox hunt is about to take place. The year is 1683, and it's a typical fair in seventeenth-century England. But what's unusual is that this fair is happening on the frozen River Thames. Welcome to the Little Ice Age!

After the balmy temperatures of the Medieval Warm Period, the Little Ice Age brought a chill to the British Isles and western Europe. It was a tough time for everyone. Crop prices went through the roof, farm animals died in the big freeze, and couriers travelling between London and the provinces were often found frozen to death, with their horses sometimes suffering the same fate. Sea ice formed around the coast and hemmed in ports. There are even reports of a polar bear harassing crofters in the Orkneys and an Inuit paddling along the Dee in Aberdeen. How reliable these stories are is open to question, but it's certain that Britain was very, very cold for many years. In 1683 – the coldest year on record – the Thames froze over for more than 10 weeks, and people held 'frost fairs' on the ice.

The Little Ice Age ended in the mid-nineteenth century, but experts disagree about when it started – some say the thirteenth century, others think 1450 was nearer the mark. Disagreements arise because the phenomenon was not simply a giant cold snap. The cooling happened at different times in different parts of the world and was interrupted by relatively warm periods. Changes in solar activity are thought to have been the cause.

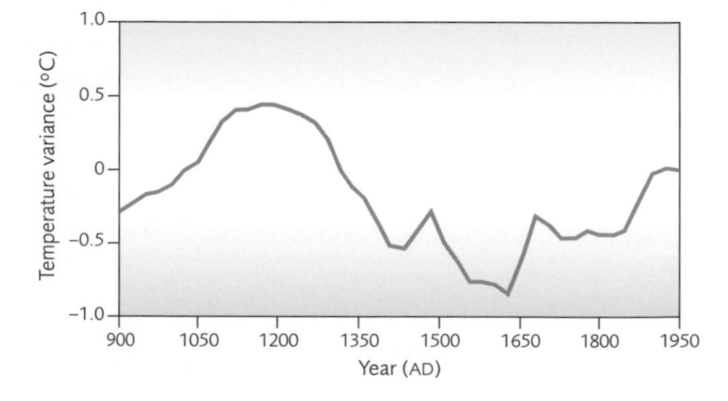

Some day, ice will return to bury the British Isles, though we're probably safe for a good few thousand years. But how likely is it that the Gulf Stream will shut down and plunge us into another Little Ice Age? Scientists have already detected an alarming 20 per cent fall in the volume of water in the Gulf Stream, and the decline has accelerated in the last few years. Despite the disturbing data, however, most people remain unconvinced. The UK Climate Impacts Programme states that 'A cooling of the UK climate over the next 100 years because of changes to the Gulf Stream is considered unlikely', although 'a weakening during the 21st century is predicted'. So perhaps it is best to sell those skis after all – at least for now!

The Future Is Wild

There's little doubt that global warming is happening. Flowers are appearing earlier; bee-eaters are breeding in England; cicadas are singing in the Baltic; and glaciers are shrinking. The evidence is all around us. Britain looks certain to change, and it seems most likely that we'll get warmer rather than cooler. The pattern looks like being summer droughts, winter rains and lots of storms.

If you look at the long and incredibly eventful history of the British Isles, the only constant is change. Our islands have changed so much and so often that global warming is simply the next agent of change. But there's a difference. For billions of years, nature alone governed the landscape. Then along came humans. At first there were only a few thousand of us, and our impact was minimal. Slowly but surely, our numbers rose, and the landscape began to change. We turned forests into fields, we divided the landscape with walls, hedges, fences and roads, we built villages that grew into towns and cities. The pace of change got faster and faster over time. In the last 50 years, fields have become mega-fields, and cities have

(LEFT) THE CAIRNGORMS COULD BECOME THE NEXT ASPEN IF GLOBAL WARMING STOPS THE GULF STREAM AND MAKES THE BRITISH ISLES COLDER.

(OPPOSITE) EVERY JANUARY IN SWITZERLAND'S ST MORITZ A GREAT HORSE RACE TAKES PLACE ON A FROZEN LAKE. THERE ARE EVEN CRICKET MATCHES HELD THERE TOO. WILL THE BRITISH ISLES SEE SUCH WINTER WONDERS?

become mega-cities. Now every square foot of the landscape bears our fingerprints, and for the first time we have begun to change our climate.

We must, as a nation, try to reduce our contribution to global warming. We have to lower the amount of greenhouse gases we release into the atmosphere. It's not just something for industry to worry about – we can help, too. Why not walk or cycle to your local shop rather than drive? Or arrange a car-share system for the school run. Turn your television and computer off rather than leaving them on standby, and unplug your mobile phone charger when you aren't using it. All these little things could really make a difference.

So those are the worries, the concerns and the darker sides. Some of them we can help to alleviate – and we have a duty to work hard at those. With others we are in the lap of the gods, and there is no point going through life being terrified of what might happen 'if things continue as they are at present'. Things seldom do, in my brief experience, and scientific extrapolation – while attempting to sound a well-meaning warning note – can often be very wide of the mark.

What we must not do, at any cost, is squander our natural heritage. We are lucky to live in a group of islands that has a rich and diverse range of wildlife, and a varied landscape that can be stunningly beautiful. Our history, both natural and cultural, is second to none. But the British Isles we all love, the British Isles we are so proud of, are also fragile. At this stage in our 3-billion-year history, only one thing is certain: that we need to put the welfare of our beloved islands ahead of our own needs so that our children and grandchildren can also boast that our land is best. If we fail them, we cannot expect their forgiveness, and these islands that are at the very heart of our being will have been sorely abused. They deserve better. They deserve the best.

Places
to visit

WHILE MAKING THE TELEVISION SERIES, I TRAVELLED THE LENGTH AND BREADTH OF THE BRITISH ISLES AND NOTICED HOW MUCH THE LANDSCAPE CHANGES IN JUST A FEW MILES, AND HOW PROFOUNDLY OUR LAND HAS BEEN INFLUENCED BY HUMAN HISTORY. THIS GAZETTEER HIGHLIGHTS SOME OF THE PLACES I VISITED, PLUS OTHERS WHERE YOU CAN FIND OUT MORE ABOUT OUR ISLANDS' JOURNEY THROUGH TIME. THE ENTRIES ARE ORGANIZED IN THE SAME ORDER AS THE BOOK'S CHAPTERS.

CHAPTER ONE

Solid Foundations

Nowhere on Earth has such a rich variety of rocks as the British Isles. During their complicated geological history, our islands' foundations travelled from near the South Pole, across the equator and through the tropics to our present location. We had ice ages, volcanic eruptions, erosion, deserts and rainforests, and we disappeared repeatedly under the sea. Because of this eventful geological past, we have 3 billion years' worth of different rocks and landscapes around us.

SCOTLAND

1 The Callanish Stones, Isle of Lewis

GRID REF. NGR NB 213330

The world-famous standing stones on the Isle of Lewis are not only a striking monument left by some of the earliest Britons – they are also made from the oldest rock in Britain. Called Precambrian gneiss, it is 3 billion years old. In fact, the Outer Hebrides are made mostly of this rock, so any road cutting or outcrop will give you a good view of its composition. You can sail to the Outer Hebrides from several places on the Scottish mainland. Daily sailings are available with Caledonian McBrayne.

Further information:
www.megalithia.com/callanish
www.calmac.co.uk

2 Torridon Mountains, Highland

GRID REF. NG900550

Between 1200 and 800 million years ago, Scotland was a hot desert. The evidence can be seen in the layered, coarse sandstone rocks that make up the magnificent Torridon Mountains in the far northwest of the country. A drive around this area will take your breath away and show you the ancient sandstones to best effect. The most westerly mountain, Beinn Alligan, is best viewed from the southern shore of Loch Torridon along the A896. The approach to Glen Torridon from Kinlochewe along this road provides spectacular views of the area. The road through the glen runs beside the Liathach mountains, which are spectacular. Beinn Eighe is best seen from the shore of Loch Clair off the B8056 (grid ref. NG770715). Slioch is an outlying peak by Loch Maree and is best seen from the Gairloch road (A832) and on leaving Kinlochewe towards Torridon (A896). The best map for this area is the Ordnance Survey Outdoor Leisure 8, 1:25000m.

Further information:
www.torridon-mountains.com

3 Fossil Grove, Glasgow

GRID REF. NS540672

About 320 million years ago, the Glasgow region was a steamy tropical swamp. Huge trees, giant dragonflies and terrifyingly large centipedes lived here until the area was submerged by water. Remains of the swamp can be seen in the remarkable Fossil Grove in Victoria Park in Glasgow. The stumps of 11 'trees' (actually giant club mosses) are preserved and protected where they were found in 1887. A fallen trunk about 26 ft (8 m) long has also survived. Fossil Grove is open daily in summer from noon to 5:00 p.m. It's in the Whiteinch area of Glasgow, and Whiteinch railway station is nearby.

Further information:
www.glasgowmuseums.com

ENGLAND

4 The Wrekin, Shropshire

GRID REF. SJ627080

This prominent, conical-shaped hill is 1335 ft (407 m) high and stands between Telford and Shrewsbury in Shropshire. Made of layers of volcanic lava and ash that formed about 670 million years ago, it is the oldest type of rock in England. The long Shropshire Way footpath goes over the Wrekin and takes in the Iron Age hillfort on the summit.

Further information:
www.mythstories.com/giantA.html (for the hill's geology, history and legends)
www.shropshireonline.gov.uk/countryside.nsf (for the Shropshire Way walk)

(5) Cow and Calf Rocks, Ilkley, West Yorkshire

GRID REF. SE113452

I had to include my own home patch of Ilkley Moor. The Cow and Calf Rocks (so called because there is a large outcrop of rock with a smaller one right next to it) are made of millstone grit, a hard type of sedimentary rock that makes up much of the Pennines. Ilkley Moor is typical millstone-grit country: rough heather moorland with brown peaty streams and tarns dotted about. If you sit quietly, you'll hear (and then see) the red grouse that live in this moorland. Ilkley has its own train station and sits on the A65 between Skipton and Leeds.

Further information:
www.information-britain.co.uk/counties.cfm?county=97

COW AND CALF ROCKS

(6) White Cliffs of Dover, Kent

GRID REF. TR 340420

The Strait of Dover separates England from France and is 20 miles (32 km) wide. The famous White Cliffs, which are more than 300 ft (90 m) high in places, have been a landmark to seafarers for thousands of years. For a good view, take the cliff path along the top. For information about the walk or the cliffs visit the new National Trust visitor centre. The White Cliffs are just east of the port of Dover, which can be reached on the A20, A2 or A256.

Further information:
www.nationaltrust.org.uk

(7) Lyme Regis, Dorset

GRID REF. SY340920

Lyme Regis and its neighbour Charmouth on the Dorset coast are famous sites for collecting Jurassic fossils (about 200 million years old). The Jurassic was the age of the dinosaurs, but you would be very lucky to find a whole skeleton. However, with patience you will definitely find ammonites, belemnites and shellfish. Fossil shops and museums in Lyme Regis and the visitor centre in the car park at Charmouth provide excellent information on how and where to look. Don't forget to read the 'Fossil Code of Conduct', which every fossil hunter must obey. It's best to visit at low tide, so call ahead to check.

Further information:
www.coastalguide.to/jurassic_coast (for Jurassic coast details)
www.amherstlodge.com/LymeRegis/fossilh.htm
For information on finding fossils in Charmouth see:
http://members.aol.com/charhercen
www.lymeregis.com/charmouth_heritage_coast_centre

WALES

(8) Mount Snowdon, Gwynedd

GRID REF SH608544

At 3560 ft (1085 m), Snowdon is the tallest mountain in Wales. It is made of 560-million-year-old rocks of various types, from mudstone and sandstone to volcanic ash. Right at the top is mudstone that formed in the sea, and preserved within it are shellfish fossils. You will see them if you look carefully in the staircase to the summit platform.

Further information:
www.hightrek.co.uk/climbing/peak/routes.htm (for hiking routes up the mountain)
http://members.aol.com/Walesrails/smr.htm (for train travel to the top)

NORTHERN IRELAND

(9) Giant's Causeway, Co. Antrim

GRID REF. C947442

The world-famous Giant's Causeway is about 20 miles (32 km) north of Coleraine and is the only World Heritage Site in Northern Ireland. Sixty million years ago, volcanic lava formed a mass of hexagonal columns that now run like stepping stones from the cliff to the sea, where they disappear under water. There are about 40,000 of the columns, the biggest of which are 43 ft (13 m) tall. Similar structures occur in Fingal's Cave and St Kilda in Scotland and are assumed to be part of the same formation. A circular walk takes you down to the Grand Causeway, past amphitheatres of stone columns and back along the clifftop. A bus service runs between the visitor centre and the Causeway.

Further information:
www.northantrim.com (for details on the geology, folklore and visitor services)
www.giantscausewayofficialguide.com

REPUBLIC OF IRELAND

(10) The Burren, Co. Clare

GRID REF. R115200

The Burren (*Boireann* in Irish) means 'rocky place'. It formed 360 million years ago when a warm tropical sea covered most of the British Isles. It is the largest limestone plateau in our islands, and it's a remarkable sight; some liken it to a moonscape that dips gently down to the sea. Deep, criss-crossing fissures called 'grykes' divide the rock into sections ('clints'), making walking difficult. The area is known for its unusual, lime loving plant life. The Burren is more than 100 square miles (260 sq km) in area and is crossed by only a few roads. Kilfenora, Lisdoonvarna and Ballyvaughan are the three largest towns in the area. There is a visitor centre at Kilfenora.

Further information:
www.theburrencentre.ie

(OVERLEAF) TORRIDON MOUNTAINS

GIANT'S CAUSEWAY

CHAPTER TWO

The Big Freeze

Nowhere in the British Isles escaped the dramatic effects of the Ice Age. Even southern England, which was never covered by ice, was fundamentally altered. At the peak of the Ice Age, a sheet of ice more than a mile thick covered northern Britain. Glaciers eroded large areas of uplands to create the characteristic landforms we see today, such as U-shaped valleys, hanging valleys and arêtes. They also left deposits of sand, gravel and rubble all over lowland areas. Because so much water was locked up in ice sheets during the Ice Age, the oceans were lower, and neither the English Channel nor the Irish Sea existed.

SCOTLAND

1 Lairig Ghru, Highland

GRID REF. NH970015

The Lairig Ghru is a U-shaped valley 2000 ft (610 m) deep that was carved out by Ice Age glaciers. It cuts through the heart of the Cairngorms and is best viewed from Aviemore station or by walking through the Charlamain Gap from Aviemore Ski Centre. The Devil's Point offers a good view from the other end. Aviemore is the main tourist centre for this area. It is well serviced by air (Inverness airport is a 40-minute drive away), rail and bus.

Further information:
www.highlandconference.com

2 Carn Mor Dearg Arête, Highland

GRID REF. NN178721

An arête is a narrow ridge between two glacial valleys. The Carn Mor Dearg arête connects the mountain of Carn Mor Dearg with Ben Nevis. If you don't mind a stiff walk, it's best seen from the top of Aonach Mor ski resort, right next to Ben Nevis (7 miles/11 km north of Fort William on the A82). Go as high as you can on the ski lifts in winter (walk in summer), and continue to the plateau above the ski runs. On a good day you'll be rewarded with a spectacular view of both the arête and Ben Nevis.

Further information:
www.nevis-range.co.uk

3 The Cuillin Ridge, Isle of Skye

GRID REF. NG446262

The Cuillins must be Britain's most spectacular mountains. The jagged peaks are arranged in a horseshoe and are connected by glacial arêtes, forming a ridge that runs all the way around the range. Walkers beware: the Cuillins are dangerous, and magnetic minerals in the rocks make compasses unreliable. Only experienced mountaineers should attempt the ridge. For a less challenging way of seeing the Cuillins, take a boat into the horseshoe for a short walk to a spectacular tarn. Trips leave daily in summer and pass close to seal colonies.

Further information:
www.skye.co.uk

4 Edinburgh Castle Rock, Edinburgh

GRID REF. NT251737

Edinburgh Castle and the Royal Mile sit on a 'crag and tail' – a glacial feature formed where a large lump of hard, resistant rock got in the way of an advancing glacier. If you look carefully at the exposed rock beside footpaths, you'll see scratches left by the glacier. The castle is in the centre of Edinburgh.

Further information:
www.edinburgh.org

5 Agassiz Rock, Edinburgh

GRID REF. NT255708

Also in Edinburgh is Agassiz Rock, an overhanging cliff. When the Swiss naturalist Louis Agassiz visited Scotland in 1840 he attributed scratches in the cliff to Ice Age glaciers. It was the first time anyone had found evidence of glaciers where there is now no ice at all. Agassiz Rock is on the south side of Blackford Hill in south Edinburgh, between the A702 and A701.

Further information:
www.scottishgeology.com/classic_sites/locations/agassizrock.html

ENGLAND

6 Norber Erratics, North Yorkshire

GRID REF. SD745690

Erratics are large rocks that have been dumped by glaciers far from their site of origin. The Norber erratics, near the village Austwick in the Yorkshire Dales, are boulders of black sandstone and shale. Glaciers carried them about half a mile from Crummockdale to the limestone plateau of the Dales. Subsequent erosion of the limestone left the harder erratics perched on small white pinnacles of limestone.

Further information:
www.yorkshire-dales.com

7 Cromer Cliffs, Norfolk

GRID REF. TG194431

The cliffs running west from the beach at Cromer are one of the best places in the British Isles to see glacial till. Till is a jumbled mass of silt, clay, sand, gravel, stones and boulders deposited by a glacier. A good spot to see erratics folded into the till is in the section of cliffs under Wood Hill. Huge lumps of chalk, hundreds of feet long, have been lifted by the ice and thrust into the till. You can find a more detailed look at the Ice Age geology of this coast on the website below.

Further information:
www.jfk.herts.sch.uk/class/geography/ks5/costal_erosion/geology_north_norfolk.htm and click on the link to 'West Runton–East Runton'

⑧ Honister Pass, Cumbria

GRID REF. NY237139

A U-shaped glacial valley in the Lake District.

Further information:
www.cumbria-the-lake-district.co.uk

⑨ Striding Edge, Cumbria

GRID REF. NY352145

A spectacular glacial arête in the Lake District that you can walk along.

Further information:
www.walkingbritain.co.uk/walks/walks1/w156.shtml

WALES

⑩ Cwm Idwal, Snowdonia

GRID REF. SH645596

A *cwm* (the Welsh name for a 'corrie' or 'cirque') is steep-walled basin high in a mountain, carved out by the top of a glacier and often containing a lake (tarn). Cwm Idwal in the Ogwen Valley in North Wales is a spectacular example – a huge natural amphitheatre with a lake in the bottom and two adjoining hanging valleys (carved by smaller tributary glaciers). It is also a National Nature Reserve and is famous for its alpine plants. One in particular is a remnant of the Ice Age: the Snowdon lily (*Lloydia serotina*). Apart from the north-facing slopes of Cwm Idwal, the only places to find it are in Arctic tundra or high in the Alps.

Further information:
www.nwt.co.uk

EDINBURGH CASTLE

⑪ Nant Ffrancon, Gwynedd

GRID REF. SH650650

A U-shaped glacial valley in Snowdonia, northwest of Cwm Idwal.

Further information:
www.countrygoer.org/snowdon.htm

⑫ Gospel Pass, Brecon Beacons, Gwent

GRID REF. SO234352

Another U-shaped valley. For details on visiting the Brecon Beacons see:
www.breconbeacons.org

⑬ Crib Goch, Gwynedd

GRID REF. SH609544

An arête in Snowdonia. There is a walk that takes in Crib Goch.

Further information:
www.go4awalk.com/walks/gwwalks/gw136.htm

NORTHERN IRELAND

⑭ Mountains of Mourne, Co. Down

GRID REF. J295248

These hauntingly beautiful mountains, with their sharp, glacier-scoured peaks, have long been an inspiration to poets. They were designated an area of outstanding natural beauty in 1986. The mountains are crisscrossed by old tracks, which make wandering through them easy. If you're feeling energetic, try the Mourne Wall – a 22-mile (35-km) track that links all the main peaks.

Further information:
www.fjiordlands.org/carlnfrd/mournes.htm
www.mournewall.freeserve.co.uk
www.geographia.com/northern-ireland

REPUBLIC OF IRELAND

⑮ Mountains of Kerry, Donegal and Connemara

The spectacular mountains of southern Ireland all show the effects of Ice Age glaciation. Good places to see glacial features include the Ring of Kerry in Co. Kerry, where there's an impressive U-shaped valley (the Gap of Dunloe); Donegal Mountains in Co. Donegal; and Connemara National Park.

Further information:
www.countykerry.com
www.irelandseye.com/aarticles/travel/nature/landscape/donegal.shtm
www.duchas.ie/en/NationalParks/ConnemaraNationalPark

GREY SEALS

Grey seals are thought to have colonized our shores during the Ice Age. The ice disappeared thousands of years ago but the seals stayed, and their white pups – whose coats provide perfect camouflage on snow and ice – now stand out conspicuously on our beaches. It's best to join a guided tour to see grey seals because disturbance can cause mothers to abandon their pups. What's more, grey seals can be aggressive and will attack and bite if you get too close. Some of the best colonies to visit are listed below.

⑯ Isle of May, Fife

GRID REF. NT654996

The Isle of May is a small island in the outer Firth of Forth, some 5 miles (8 km) east of Crail. The seals here breed in autumn. No scheduled ferries sail in the winter so you will have to charter your own boat from Crail.

Further information:
Scottish Natural Heritage office, Cupar, Fife. Telephone 01334 54038 .

⑰ Donna Nook, Lincolnshire

GRID REF. TF425992

You can see the seals from the car park at this National Nature Reserve. A warden is usually on site to provide information and make sure nobody strays from the path. There's an RAF bombing range nearby, but it doesn't seem to bother the seals.

Further information:
www.english-nature.org.uk
www.lincstrust.org.uk

⑱ Cardigan Island Farm Park, Gwbert, near Cardigan

GRID REF. SN163497

A colony of seals lives in the caves below Cardigan Island Farm Park, and they can often be seen basking on the rocks. They give birth in September–October, when the fluffy white pups can also be seen.

Further information:
www.cardiganshirecoastandcountry.com/cardiganisland.htm

⑲ Ramsay Island, Pembrokeshire

GRID REF. SM700235

The most important colony of grey seals in Wales is on the northern tip of St Bride's Bay. There are daily boats from St Justinian's between June and September.

Further information:
www.visitpembrokeshire.co.uk (for tourism details)

CHAPTER THREE

After the Ice

When the ice receded at the end of the Ice Age, sea levels rose and cut Britain off from Europe. The newly formed coasts, estuaries and offshore islands became rich habitats for all kinds of new arrivals, and forests flourished on the mainland. The loss of the weight of ice in northern Britain caused the whole of the country to rebound and tilt, a process that continues to this day. As a result, Scotland is slowly rising and southern England is slowly sinking (though only by a millimetre or so a year).

SCOTLAND

1 Jura Raised Beach, Hebrides

GRID REF. NR529874

The west coast of Jura is the best place in the British Isles to see raised beaches, perhaps because the ice sheets were at their thickest near here in the Ice Age. Some beaches are now 130 ft (40 m) above sea level. You'll need to be a keen walker to see them – Jura has only one road and it runs up the east coast.

Further information:
www.scotland-inverness.co.uk/jura.htm

2 Abernethy Forest, Tayside

GRID REF. NH975168

Abernethy Forest in the Cairngorms National Park is the last remnant of the ancient pine forest that covered northern Britain shortly after the Ice Age. It is also the largest stretch of native pine forest in the British Isles. Huge, knotty Scots pine stand among juniper, dwarf birch and rowan, and the understorey has primroses, bluebells, wood sorrel and anemones. You may also see some of the distinctive birds that have evolved to live in northern pine forests, including common and Scottish crossbills, black grouse, capercaillies, siskins and crested tits. Abernethy is just off the B970 between Aviemore and Boat of Garten.

Further information:
www.nethybridge.com/html/thingstodo/explore.htm

3 Ythan Estuary and Forvie Nature Reserve, Grampian

GRID REF. NK020270

This Site of Special Scientific Interest is 12 miles (20 km) north of Aberdeen. It is a tranquil inlet with a sandy shore, 19 miles (30 km) of sand dunes, mud flats, and mussel beds, all of which are rich in life. The sands formed at the end of the Ice Age some 10,000 years ago, when vast quantities of sediment were transported by rivers from the melting ice to the coast and deposited offshore. When the sea level rose, the sediments were washed back on to the beach to form the dunes we see now. It is a wonderfully peaceful place, and you can be guaranteed to see a wide range of wildlife. Eider ducks are especially common – about 6000 come here to breed in spring, and about 1000 stay through the winter.

Further information:
www.nnr-scotland.org.uk/reserve.asp?NNRId=23
www.boddam.demon.co.uk/mpage5.htm

ENGLAND

4 Newtondale, North Yorkshire

GRID REF. SE808915

Newtondale is a steep-sided river valley in the North York Moors that was carved out of the land by a catastrophic flood of glacial meltwater 13,000 years ago. It is 8 miles (13 km) north of Pickering and can be seen from the A169. The North York Moors railway runs through the valley and provides a good way of seeing it in its entirety.

Further information:
www.nymr.demon.co.uk
(for train times)

5 Bempton Cliffs, East Yorkshire

GRID REF. TA197741

Bempton Cliffs is an RSPB seabird reserve on the Yorkshire coast. The power of the sea, the dramatic chalk cliffs and the noise of the birds make this a spectacle not to be missed. It is the second-largest seabird colony and the only gannetry on the British mainland. The cliffs run for 5 miles (8 km), and although the reserve is only a few yards wide, 220

ABERNETHY FOREST

species of plants have been recorded. Erosion of the soft rock has created narrow ledges and cracks for 60,000 pairs of kittiwakes. There are also fulmars, guillemots, razorbills, puffins and gannets. Bempton Cliffs are off the B1229, 4 miles (6.5 km) northwest of Flamborough Head.

Further information:
www.rspb.org.uk/reserves/guide/b/bemptoncliffs

6 Snettisham RSPB Reserve, Norfolk

GRID REF. TF647335

The Wash, Britain's largest estuary, is a vital habitat for wading birds, ducks and geese. The mud provides food all year round, as well as roosting sites, particularly in winter. Snettisham on the north Norfolk coast is the place to see one of our greatest wildlife spectacles, when roosting knot and other waders are forced off the mud by the rising tide and fly en masse to a nearby salt marsh. Snettisham is open all year round, but for the best times to visit see below.

Further information:
www.rspb.org.uk/reserves/guide/s/snettisham

7 Tresco Gardens, Isles of Scilly

GRID REF. SV900150

Since the Ice Age ended, the warm waters of the Gulf Stream have bathed our islands, giving us a damp but very mild climate compared to continental Europe. One of the warmest places in Britain is the Isles of Scilly, about 28 miles (45 km) south of Land's End. Here the climate is frost free, and subtropical plants grow in magnificent gardens. The best known is the Abbey Garden on the island of Tresco. Plants from Australia, Borneo, South Africa and Brazil grow side by side in the open air. The garden is open all week from 9:30 a.m. to 4:00 p.m.

TRESCO GARDENS

Further information:
www.tresco.co.uk
www.scillyonline.co.uk

WALES

8 Borth Tree Stumps, Dyfed

GRID REF. SN606897

The popular seaside resort of Borth in Wales, about 5 miles (8 km) north of Aberystwyth, has a special visitor attraction: a whole forest on the beach! At low tide you can see the remains of ancient conifer trees that were submerged by the rising sea at the end of the Ice Age. You'll need to look carefully because the dark brown stumps are often covered with seaweed.

Further information:
www.dyfi.com/html/borth.html

There are also fossil forests offshore near Skegness, in the Severn Estuary, and in the Solent near Southampton.

9 Llangorse Lake, Powys

GRID REF. SO128271

Llangorse Lake in the Brecon Beacons National Park is the largest natural lake in South Wales, with a circumference of 5 miles (8 km) and a total area of 327 acres (132 hectares). It lies in a shallow basin that was carved out by an Ice Age glacier, and dammed by glacial gravel. It is 8 miles (13 km) from Brecon on a minor road off the A40, and there's a car park to the west of Llangorse village on the northern shore.

Further information:
www.brecon-beacons.com/llangorse-lake.htm

REPUBLIC OF IRELAND

10 Peatland World, Rathangan, Co. Kildare

GRID REF. N271226

When the ice melted at the end of the Ice Age, it left behind pools that filled with dead vegetation and formed peat bogs in much of central Ireland. Peatland World is in the heart of the Bog of Allen, a raised peat bog that covers parts of counties Kildare, Laois and Offaly. It is an interpretation centre that provides information on the formation and use of peat through the ages.

Further information:
http://kildare.ie/touristguide/ThingsToDo/moreinfo.asp?RecordID=86

11 Esker Riada, Co. Offaly

GRID REF. N201231

Melting ice sheets left gravel, sand and rocks in mounds and ridges all over central Ireland. Long, winding ridges of glacial gravel ('eskers') were reputedly used by early Christian settlers as highways. One famous example is the Esker Riada near Tullamore in Co. Offaly, right in the centre of Ireland, which was supposedly part of the highway from Dublin to Galway.

Further information:
www.offaly.ie/visitoffaly.asp

Another place to see eskers is the area around Carstairs Kames (grid ref. NS950450), 25 miles (40 km) southeast of Glasgow

CHAPTER FOUR

Taming the Wild

The arrival of agriculture led to a transformation of the British Isles. The great wildwood that once covered the country was slowly cleared to make way for crops, and people began to live in larger, more settled communities. The earliest farmers left few signs of their presence apart from stone monuments. In later times, such as the Roman and medieval periods, farmers left an increasing mark on the landscape.

SCOTLAND

(1) Skara Brae, Orkney Mainland

GRID REF. HY231187

Stone Age villages are as rare as hen's teeth, and that's what makes Skara Brae so remarkable. A complete Neolithic village, it was built about 5000 years ago on the main island in the Orkneys. You can see beds, chairs, fireplaces and even shelves that may have been used to display ornaments. The village was discovered when sand dunes shifted and exposed the stone buildings.

Further information:
www.orkneyjar.com/history/skarabrae
www.visitorkney.com

ENGLAND

(2) Hadrian's Wall, Tyne & Wear to Cumbria

GRID REF. NY790688

Hadrian's Wall – the northernmost limit of the Roman Empire – was built to defend England from marauding Scottish barbarians (though there is little evidence of conflict). It once ran from coast to coast but much has now gone. The remains are still an impressive testimony of Roman power, and wonderfully evocative to visit, making you wonder what life must have been like for the legions of Roman soldiers who were stationed here. The National Trust owns 6 miles (10 km) of the wall and runs a visitor centre at Housesteads Fort, the best preserved fort on the Wall. The fort is northeast of Bardon Mill on the B6318 (grid ref. NY790690).

Further information:
www.nationaltrust.org.uk
www.hadrians-wall.org

(3) Carn Euny, Cornwall

GRID REF. SW402288

This ancient village is thought to have Stone Age origins, but most of what you see dates back to about 100 BC. The main attraction of the site is a mysterious underground passage.

Further information:
www.st.just.online.freeuk.com/carn-euny.htm
www.stonepages.com/england/carneuny.html

(4) Chysauster, Cornwall

GRID REF. SW473350

A Stone Age village dating to about 2000 BC.

Further information:
www.cornwall-online.co.uk/english-heritage/chysauster.htm
www.stonepages.com/england/chysauster.html

(5) Wharram Percy, North Yorkshire

GRID REF. SE860646

Remains of medieval villages are not uncommon in the British Isles, but few are as well documented and excavated as this one in Yorkshire. Wharram Percy was built in the tenth century and abandoned in 1517, when residents were evicted to make way for sheep pasture. There isn't much to see besides a ruined church and some grassy lanes and old foundations, but the site has been well excavated and is managed by English Heritage. It's 6 miles (10 km) southeast of Malton on a minor road off the B1248, half a mile south of Wharram le Street. Admission is free and it's open all year round.

Further information:
www.english-heritage.org.uk

(6) Stonehenge, Wiltshire

GRID REF. SU120422

This amazing stone circle is a World Heritage Site and one of the most evocative and mysterious places in the British Isles. No one knows why it was built. The earliest parts are more than 5000 years old, making them older than the pyramids of Giza. Stonehenge is managed by English Heritage and is surrounded by National Trust land. It is 2 miles (3 km) west of Amesbury in Wiltshire.

Further information:
www.english-heritage.org.uk

(7) Epping Forest, Essex

GRID REF. TQ814970

This 6000-acre (2400-hectare) forest to the north of London is all that remains of an ancient woodland that once covered most of Essex. In medieval times the trees were repeatedly cut back

(pollarded) to produce poles for making fences and charcoal. Now the trees have a peculiar shape and thick canopy. The forest is an important habitat for invertebrates, roe deer and fallow deer.

Further information:
www.touruk.co.uk/london_parks_gardens/eppingforest_park1.htm

⑧ Windsor Great Park, Berkshire

GRID REF. TQ023784

Ancient woodland is defined as an area that has been continuously forested since 1600. There is very little of it left in Britain, but Windsor Great Park has the largest collection of ancient oaks in the country, including many that are more than 600 years old, and one estimated to be 1300 years old. Thanks to their great age, the trees have acquired some unusual inhabitants, including very rare species of beetles and spiders. The park is a mixture of woodland and grassland, with a wide variety of trees, including beech, oak, sweet chestnut, birch and conifers. Near Windsor in Berkshire, it is open all year from dawn till dusk and admission is free.

Further information:
www.thamesweb.co.uk/windsor/info/grtpk.html

Other ancient British woodlands include the New Forest (www.thenewforest.co.uk), sandwiched between Bournemouth and Southampton in Hampshire, and the Wye Valley on the Welsh border near Chepstow (www.wyevalleyaonb.co.uk).

⑨ Thurne Dyke Mill, Norfolk

GRID REF. TG401159

The vast wetlands of East Anglia were drained in the seventeenth and eighteenth centuries to create farmland. Windmills, such as the one at Thurne Dyke, were used to pump water out of the fields to keep them dry. Most of the windmills have disappeared, but there are still a few in working order. Thurne Dyke Mill is near the village of Thurne in the Norfolk Broads and is open daily from Easter to the end of September, between 9:00 a.m. and 8:00 p.m. An exhibition tells you more about the history of the Broads.

Further information:
www.norfolkwindmills.co.uk

NEWGRANGE

⑩ Grimes Graves, Norfolk

GRID REF. TL818898

Stone Age miners dug these extraordinary pits some 4000 years ago to find flint for making tools. They used animals' shoulder blades as spades to dig around 350 pits, and deer antlers to prise flint 'nodules' from the chalk bedrock. It's a strange experience climbing 30 ft (9 m) down a ladder and seeing the prehistoric mine-shafts around you – a rare glimpse of an ancient industry. Many of the flints were used to make the stone axes that contributed to Britain's deforestation after the Ice Age. Grimes Graves is 7 miles (11 km) northwest of Thetford, off the A134, and is managed by English Heritage.

There are also Neolithic flint mines at Cissbury and Harrow Hill in Sussex.

Further information:
www.britainexpress.com/articles/Ancient_Britain/grimes-graves.htm
www.brecks.org.uk/exploref/pop-uppa/grimesgr.htm
www.findon.com/cissbury/cissbury.htm#flint_mines
www2.prestel.co.uk/aspen/sussex/harrowhill.html#arch2

⑪ The Ridgeway, between Avebury, Wiltshire, and Uffington, Oxon

GRID REF. SU299860

This ancient road runs for 85 miles (137 km) and passes along a natural chalk ridge. For as long as 6000 years, it was used as a highway by traders, drovers, armies and invaders alike. There are long barrows, standing stones and hillforts along the route.

Further information:
www.nationaltrails.gov.uk/ridgewayframeset.htm
www.singlefile.uku.co.uk/the_ridgeway.htm

WALES

⑫ Chepstow Castle, Gwent

GRID REF. ST550950

Said to be the oldest stone castle in Britain, this is a magnificent reminder of the power of the Normans. It was built in 1067 and overlooks the River Wye in Chepstow on the Welsh border, a few miles from the old Severn Bridge.

Further information:
www.castlewales.com/chepstow.html

REPUBLIC OF IRELAND

⑬ Newgrange, Co. Meath

GRID REF. O301274

This extraordinary burial chamber was erected by Stone Age farmers just over 5000 years ago, at about the same time that Stonehenge was built. It sits in fertile agricultural land in Co. Meath, about 50 miles (80 km) north of Dublin. It is a huge, circular, grass-covered mound, not unlike the Tellytubbies dome! Strange carvings cover the stones, and a long passage takes you down into the central burial chamber. It is well worth a visit and guided tour. The site is open all year, but can be accessed only from the visitor centre.

Further information:
www.heritageireland.ie/en/HistoricSites/East/BrunaBoinneVistorCentreNewgrangeandKnowthMeath
www.knowth.com/newgrange.htm

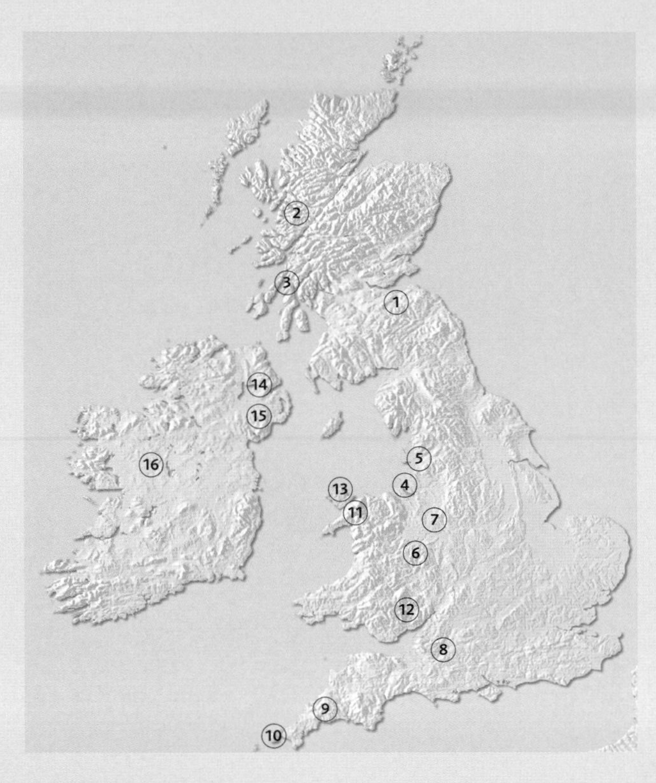

CHAPTER FIVE

Forging Ahead

The Industrial Revolution brought canals, railways, coal mines and factories to Britain, as well as a massive increase in the population. Towns expanded as people flooded in from the countryside in search of work, but conditions were often dirty and overcrowded. The new industries brought great wealth, allowing the upper classes to build opulent country homes with landscaped gardens.

SCOTLAND

1 Lady Victoria Coal Mine, Lothian

GRID REF. NT350650

This colliery is a fine example of Victorian engineering. During its working life it produced 40 million tons of coal and employed 2000 people. The visitor centre has interactive displays and a mock coalface so that you can experience what it was like for the miners under ground. Located at Newtongrange, 9 miles (14 km) southeast of Edinburgh.

Further information:
www.scottishminingmuseum.com/welcome.html

2 Fort William to Mallaig Railway, Highland

GRID REF. NN105735

A great way to spend a day in the Scottish Highlands is to take the 84-mile (135-km) round trip from Fort William to Mallaig and back again on a steam train. Not only does this journey take you through some of the most spectacular countryside in the British Isles, it also takes you across the breathtaking 1248-ft (380-m) long Glenfinnan Viaduct, which opened in 1901.

Further information:
www.steamtrain.info
www.nas.gov.uk/miniframe/edu_primary_resources/l_middle_home.htm

3 Crinan Canal, Argyll

GRID REF. NR785945

The 9-mile (14-km) long Crinan Canal opened in 1801 and dramatically improved transport between the Clyde Estuary and Scotland's west coast and islands by creating a shortcut across the Kintyre peninsula. Until the railways were built, the fastest way to get from Glasgow to Inverness was by steamer, using the Crinan Canal and the Caledonian Canal, and usually calling at Oban en route. The Crinan Canal runs through beautiful scenery and is a reminder of times when life moved at a slower, gentler pace.

Further information:
www.undiscoveredscotland.co.uk/crinan/crinancanal

ENGLAND

4 Albert Docks, Liverpool

GRID REF. 341895

A wander around these impressive dockyards will take you back to the middle of the nineteenth century, a time when Liverpool grew rich on trade and shipping. The Albert Docks were built to accommodate sailing ships; when steam took over, the docks were too shallow and fell out of use. They were beautifully restored in the 1980s and now have a fascinating museum.

Further information:
www.merseyworld.com/albert

The Mersey Maritime Museum is on the north side of the docks, near Strand Street. It is open all year and admission is free.
www.liverpoolmuseums.org.uk/maritime/index.asp

5 Wigan Flashes, Lancashire

GRID REF. SD585302

Wigan's 'Flashes' are lakes and wetlands formed from subsided mines. In the heart of urban Wigan, the 593-acre (240-hectare) site shows just how successfully industrial wasteland can be turned into a valuable wildlife reserve. Now managed by the Lancashire Wildlife Trust and the RSPB, the Flashes include two Sites of Special Scientific Interest. Bitterns overwinter in the reed beds here.

Further information:
www.british-publishing.com/Pages/wiganBG/WildSide.html
www.wildlifetrust.org.uk/lancashire

6 Ironbridge Gorge Museums, Shropshire

GRID REF. SJ674035

Ironbridge was the birthplace of the Industrial Revolution and is now a World Heritage Site. There are 10 museums and a spectacular iron bridge spanning the gorge – the world's first cast-iron bridge. You can also see iron smelting, carpentry, candle making, printing and many other skills in action. Ironbridge is 7 miles (11 km) from Telford and is open all year.

Further information:
www.ironbridge.org.uk

7 Gladstone Pottery Museum, Stoke-on-Trent, Staffordshire

GRID REF. 912432

Stoke-on-Trent was the centre of Victorian Britain's thriving pottery industry. The Gladstone Pottery Museum in Longton is a beautifully preserved Victorian pottery, complete with bottle ovens and a cobbled courtyard. It is very 'hands-on' – you can throw

GLADSTONE POTTERY MUSEUM

a pot, make a china flower and have a go at decorating a plate. You can also find out what a jolleyer, a jigger and a saggar maker's bottom knocker used to do! Gladstone Museum is open all year.

Further information:
www2002.stoke.gov.uk/museums/gladstone

8 Kennet and Avon Canal between Bath, Avon, and Reading, Berkshire

GRID REF. SU005615

A good way to spend a holiday is to hire a narrowboat and sail up the Kennet and Avon Canal, which runs for 87 miles (140 km) between Bath and Reading. It was built in 1810 and passes through some of the most beautiful countryside in England. One of the highlights is the Caen Hill locks – a 'flight' of 16 locks that takes the canal up a steep hill through Devizes.

Further information:
www.katrust.org/canal.htm
www.katrust.org/caenhillflight.htm

9 Wheal Martyn Clay Museum, Cornwall

GRID REF. SX005555

The fine clay used in the pottery industry came from Cornwall, and the Wheal Martyn Museum is an

CANAL HIGHLIGHTS

In the mid-nineteenth century Britain had about 4000 miles (6400 km) of busy canals, which were used for transporting all sorts of goods, from iron ore to coal and pottery. Roads and railways made canals obsolete, and the canal network gradually declined during the twentieth century. Today about 2000 miles (3200 km) of canals are still in use, and they provide a great way of seeing both the countryside and the ingenuity of the Victorian engineers and builders who built the network. Some of the highlights of Britain's canals are listed below.

BEST RISE: Devizes locks on the Kennet and Avon Canal. There are 29 locks in a short stretch, including a 'flight' of 16 locks up Caen Hill.

SCARIEST AQUEDUCT: Pontcysyllte Aqueduct near Llangollen in North Wales. Built by William Jessop and Thomas Telford and opened in 1805, this high-rise canal passes over the River Dee on stone pillars 125 ft (38 m) tall.

BEST LIFT: Anderton boat lift on the Trent Mersey Canal. It was designed by Edwin Clark and opened in 1875.

BEST WILDLIFE: Kennet and Avon Canal's water voles. After being decimated by feral mink, these rare aquatic mammals are making a comeback in some places thanks to conservation efforts.

SPOOKIEST TUNNEL: Standedge Tunnel on the Huddersfield Narrow Canal is more than 3 miles (5 km) long, making it the longest tunnel on the canal network. It was built by Benjamin Outram and John Ruth and opened in 1801.

BEST SWING: Barton Swing Aqueduct on the Bridgewater Canal was built by James Brindley and opened in 1761.

BEST CROSSING: The Caledonian Canal connects Loch Dochfour, Loch Ness, Loch Oich and Loch Lochy, forming a shipping route between Scotland's west and east coasts. The canal includes 29 locks and four aqueducts.

BEST CATCH: Pike. The deep water of the Exeter Ship Canal is ideal for pike, which can reach 20 lb (9 kg) in weight here.

WELSH SLATE MUSEUM

excellent place to discover more about the geological origins of clay and how it is used. There are three trails to follow and a 'pit view', where you can watch a working clay pit in action. Wheal Martyn is in Carthew, 2 miles (3 km) north of St Austell on the B3274 (look for the brown signs at the Stenalees roundabout).

Further information:
www.wheal-martyn.com

(10) Geevor Tin Mine and Levant Mine, Cornwall

GRID REF. SW367344

Just outside the village of Pendeen, 7 miles (11 km) west of Penzance, is the Geevor Tin Mine. Until 1990 it was an active mine, but now it is solely a museum. The setting is beautiful – the museum is perched on cliffs overlooking the sea on the western tip of Cornwall.

Just half a mile (800 m) away is the Levant Mine, a tin and copper mine owned by the National Trust. The Levant Mine has a working steam-powered 'beam engine', which was used to pump water out of tunnels below sea level.

Further information:
www.geevor.com
http://freespace.virgin.net/levant.mine

WALES

(11) Welsh Slate Museum, Llanberis, Gwynedd

GRID REF. SH590601

This museum is on the shore of Llyn Padarn on the flank of Snowdon in North Wales. It provides a fascinating look at the Welsh slate industry, which stretches back to the early nineteenth century; Welsh slate is said to have 'roofed the world' during the Industrial Revolution. The quarries form great caverns in the mountain, and there are many smaller, abandoned quarries in the area (such as Cedryn quarry in the Conwy Valley in Cwm Eigiau, which operated between 1827 and 1868).

Further information:
www.nmgw.ac.uk/wsm
www.penmorfa.com/Slate

(12) National Mining Museum, Blaenafon, Gwent

GRID REF. SO238087

This coal mine-turned-museum near the Brecon Beacons in South Wales is a good place to find out what life was like at the coalface. You can take a lift 300 ft (90 m) down into the mine for a guided underground tour with a working miner. The museum has interactive displays, and the colliery buildings are open to the public. The museum is open from February to November.

Further information:
www.nmgw.ac.uk/bigpit

(13) Parys Mountain, Anglesey

GRID REF. SH436899

This copper mine in Anglesey has been used since the Bronze Age. In the eighteenth century the metal was used to make coins and to sheathe battleships so that barnacles couldn't grow on them. It's a remarkable site with a long history, and it was one of the contenders on the BBC programme *Restoration*. There's no public access under ground, but you can walk around the mountain.

Further information:
www.angleseymining.co.uk/ParysMountain/HomeParys.htm
www.bbc.co.uk/wales/historyhunters/locations/pages/1_3_parys_mountain.shtml

NORTHERN IRELAND

(14) Patterson's Spade Mill, Templepatrick, Co. Antrim

GRID REF. J284842

If you think a spade is just a spade, think again. Spades have been, and still are, very important in Ireland, and there are a staggering 171 different designs. Patterson's Spade Mill is owned by the National Trust and is the last working water-driven spade mill in Britain. You can see the whole process of spade making from beginning to end – a rural industry that has almost vanished. You can also order your own made-to-measure spade.

Further information:
www.ntni.org.uk/places/placestovisit.cfm?id=15

(15) Irish Linen Centre, Co. Antrim

GRID REF. J267644

The history of the linen industry in Ireland is told in this museum in Lisburn. It includes exhibitions and hands-on demonstrations.

Further information:
www.2hwy.com/ir/i/ihenluum.htm

REPUBLIC OF IRELAND

(16) Famine Museum, Strokestown Co. Roscommon

GRID REF. M193281

The greatest disaster to strike industrial Europe was the Irish potato famine of the 1840s, when more than a million people died of starvation. The Famine Museum is in Co. Roscommon, 90 miles (145 km) west of Dublin.

Further information:
www.geocities.com/willboyne/nosurrender/FamMuseum.html

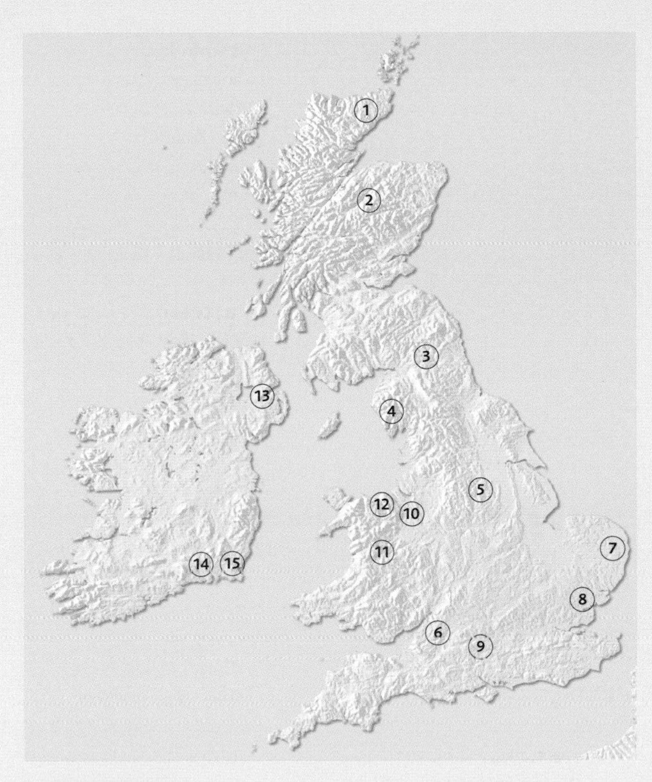

CHAPTER SIX

The Modern Era

Nowhere in the British Isles is pristine wilderness – every square inch has been affected in some way by people, especially in the last 200 years. All our activities, industrial, urban and recreational, have taken their toll on the environment, and a lot of time and money is now being spent on either limiting the damage or trying to put right past wrongs. In many ways Britain and Ireland are heartening places to live. Go anywhere and you will find organizations and individuals working hard to make our environment better. Here are just a few examples.

SCOTLAND

1 Forsinard Nature Reserve, Sutherland

GRID REF. NC891439

The Flow Country is the largest remaining area of blanket bog in the world, covering almost 26,000 acres (10,000 hectares). Much of the area was damaged by forestry, which lowered the water table and dried out the wetlands. Forests also fragmented the open landscape that is so important for ground-nesting birds. Thanks to the RSPB and others, the tax incentives that promoted forestry were stopped and the area is now undergoing intensive restoration. There's a visitor centre on the old railway line at Forsinard, and you can go on walks from there, but access to the area is limited because of the conservation projects.

Further information:
www.scotlandindex.net/rspb_main.htm

2 Loch Garten, Highland

GRID REF. NH989162

Ospreys were systematically persecuted and became extinct as a breeding species in England in 1840 and in Scotland in 1916. In the 1950s they reappeared, and there are now around 160 pairs. Loch Garten is one of the few places in which you can watch these magnificent birds. The loch is in Abernethy forest about 8 miles (13 km) from Aviemore.

Further information:
www.rspb.org.uk/scotland/action/50years.asp
www.undilutedscotland.com/htm_docs/ospreynesting.htm
www.ospreys.org.uk/AWOP/About%20Ospreys.htm

You can also see breeding ospreys in Rutland Water, Leicestershire.
www.ospreys.org.uk/AWOP/Home.htm

FLOW COUNTRY

ENGLAND

3 Kielderhead, Northumberland

GRID REF. NT 662004

These 10,000 acres (4000 hectares) of moorland are a sharp contrast to Kielder Forest next door. Kielderhead is an area of blanket bog, heather moorland and woodland, and it is an excellent place for bird-watching. It is managed by the Forestry Commission and the Northumberland Wildlife Trust.

Further information:
www.english-nature.org.uk/special/nnr/nnr_details.asp?NNR_ID=206

4 Grizedale Forest, Cumbria

GRID REF. SD334950

Just east of Coniston Water is a very good example of a Forestry Commission forest that combines economic forestry with recreation and conservation. Instead of dense rows of conifers, there's a forest theatre, sculpture trail, forest classroom, high-level canopy walk, mountain biking and walking trails.

Further information:
www.cumbrialakedistrict.com/grizedale.htm

5 Five Weirs Walk, Sheffield

GRID REF. SK375890

This fascinating and cheering walk runs for 5 miles (8 km) through the heart of industrial Sheffield. It combines the real grit of Victorian industry with regeneration and wildlife. Because this city was so heavily industrialized, the River

Don and its banks were almost completely dead. Now, with the help of legislation to clean up industry and hard graft by the Five Weirs Walk Trust, it has been transformed. Birds, insects and fish are returning (you are bound to see a fisherman on your walk).
Further information:
www.fiveweirs.co.uk

This website also tells you where to get a booklet about the history and future of the River Don.

⑥ Cotswold Water Park, Gloucestershire

GRID REF. SU020940

About 5 miles (8 km) east of Cirencester is a large area of waterways created by gravel extraction, with more than 130 lakes covering 40 square miles (104 sq km). There's a range of water sports to try, and wildlife abounds in the quieter areas.
Further information:
www.waterpark.org

⑦ Norfolk Broads

GRID REF. TG347180

The Broads were excavated by peat diggers in medieval times, and the waterways they created are among the most popular holiday spots in the country. The area is a haven for wetland wildlife and is home to many unusual butterflies and plants. However, it requires active management to keep the water flowing and stop it filling with silt or being smothered by scrub.
Further information:
www.broads-authority.gov.uk/broads/indexie.html
www.norfolkbroads.com/guide/histnorf.htm

⑧ Abbotts Hall Farm, Essex

GRID REF. TL963146

In 1991 Essex Wildlife Trust bought this 700-acre (280-hectare) arable farm and began turning it into a showpiece to demonstrate how modern farming and wildlife can coexist. By digging ditches, planting hedgerows and copses, and farming organically, the conservationists have regenerated the area and brought back wildlife. They are also knocking holes into the sea wall to allow the sea to flood in, turning farmland back into salt marsh.
Further information:
www.essexwt.org.uk/Sites/Abbotts%20Hall.htm

⑨ Greenham Common, Berkshire

GRID REF. SU492648

Greenham Common near Newbury is etched into people's minds as a cruise-missile base, but the Greenham Common Trust is now turning this symbol of the Cold War back into heathland. Since 2000 there has been unrestricted public access, and you can see for yourself how tarmac and concrete are slowly being reclaimed by orchids, adders, deer and nightingales.
Further information:
www.greenham-common-trust.co.uk

WALES

⑩ Llangollen Canal, between Nantwich, Cheshire, and Llangollen, Clwyd

GRID REF. SJ215415

This canal runs for 41 miles (66 km) through some of the most beautiful countryside in England and Wales. It is a good example of how these old arteries of Victorian Britain have become a peaceful sanctuary for both wildlife and humans. Boating is big business for the holiday trade, and the canal sides and water are kept clear by dredging, allowing fish, birds, mammals and invertebrates to thrive. Water voles are also finding canals a real haven. Don't miss the spectacular Pontcysyllte Aqueduct (*see* box 'Canal Highlights', p. 203).
Further information:
www.canaljunction.com/cllang.htm

⑪ Lake Vyrnwy, Powys

GRID REF. SH989211

This artificial lake, created by a dam built in 1880, drowned the village of Llanwddyn. In dry summers, when the water level drops, you can sometimes see the remains of the village sticking out of the water. The site is managed by the RSPB and Severn Trent Water, and the whole area around the reservoir is worth exploring. There's a large patch of heather moorland that is actively managed to help breeding birds; sessile oak woodland; and around 5000 acres (2000 hectares) of commercial forestry.
Further information:
www.rspb.vyrnwy.org/eng_map1.htm

⑫ Brickfields Pond, Rhyl, Clwyd

GRID REF. SJ012802

This small but lovely nature reserve started life as a flooded clay pit and became a popular spot for fly-tipping. Then Denbighshire Countryside Service carried out a series of successful restoration projects. Now you can enjoy bird-watching, fishing and walking, or you can sit in the glass-fronted community room and watch the watery wildlife.
Further information:
www.denbighshire.gov.uk/www/dccportal.nsf

NORTHERN IRELAND

⑬ Belfast Lough Reserve, between Co. Antrim and Co. Down

GRID REF. J370770

Most of the mud flats in the Belfast Lough have been reclaimed for industry, but one area of mud and lagoon is an RSPB reserve and an important feeding place for waders, ducks and geese. This is a great place to see wildlife right in the heart of major industry.
Further information:
www.rspb.org.uk/reserves/guide/b/belfastlough

REPUBLIC OF IRELAND

⑭ Fenor Bog, Co. Waterford

GRID REF. J253102

Just 10 miles (16 km) from Waterford City is Fenor Bog, a 32-acre (13-hectare) area of swampy grassland and herbaceous plants. Until the 1920s Fenor Bog was dug for peat, but now the area is being regenerated to provide a habitat for wetland plants and invertebrates. Work began recently (the site was purchased in 1999), so this is a good place to see conservation in action.
Further information:
www.ipcc.ie/sitefenorwater3.html
www.ireland.travel.ie

⑮ Wexford Wildlife Reserve, Co. Wexford

GRID REF. T307124

This is Ireland's most important site for wildfowl, with 236 recorded bird species, including 37 categorized as scarce. The reserve is in Wexford Harbour, where there are large areas of sand bars, mud banks and shallow water. The 'slobs' (a local name for reclaimed land) to the north and south of the harbour are also important wildlife havens. Several pools have been created to encourage wildfowl into the area.
Further information:
www.southeastireland.com/detail.asp?memberID=466

GRASSHOPPER

CONSERVATION ORGANIZATIONS

Bat Conservation Trust

(www.bats.org.uk)
Ever wondered how a bat detector works or how to tell the difference between a pipistrelle and a lesser horseshoe? If so, this trust is for you. It offers a wide range of bat-related training courses to help a network of volunteers monitor UK bat populations. Members campaign locally, nationally and internationally for bat conservation, and encourage everyone to appreciate bats – the stars of the night.

British Trust for Ornithology

(www.bto.org)
The BTO organizes volunteer surveys of birds, so if you're a bird-watcher, you can get involved. You can take part in Garden Birdwatch or collect nest records, and anyone over the age of 14 can participate in bird ringing under the supervision of a qualified ringer. Contact the regional representative in your area for further details.

Buglife

(www.buglife.org.uk)
According to some estimates, more than two-thirds of all species on the planet are invertebrates, and this is the first organization in Europe committed to their conservation. Species such as the short-haired bumblebee have become extinct in Britain in the last 15 years. Buglife aims to prevent further extinctions by promoting research and management of land and water to enhance invertebrate biodiversity. If you want to help, check out its website to find out how to encourage more bees and other creepy crawlies into your garden.

Butterfly Conservation

(www.butterfly-conservation.org)
The British Isles have a profusion of beautiful native butterflies and moths, and this organization aims to protect them from the threat of habitat destruction. There are many ways you can help, such as by joining a local branch or by helping with surveys and monitoring programmes. You can also make your garden more butterfly-friendly by following the organization's advice. If you need an excuse to sit in your garden on a sunny day, why not record your butterfly sightings and send the records to the Butterflies for the New Millennium project? See the website for more details.

Mammal Society

(www.abdn.ac.uk/mammal)
Working to protect British mammals, the Mammal Society carries out a number of surveys every year. It's keen to involve volunteers in recording mammal sightings – and don't worry if you can't tell the difference between a common and an edible dormouse because it also runs training workshops at weekends. Its 'National Survey of Road Deaths' records how many wildlife casualties occur on our roads. This vital information can identify hotspots and help inform government bodies on future road-building plans.

Plant Life

(www.plantlife.org.uk)
This charity is dedicated to conserving plants in their natural habitats, both in the UK and internationally. The charity is perfect for budding botanists. It runs a range of surveys catering for people of varying botanical skills, so you don't have to be an expert to help. The surveys aim to record changes in the distribution of plant species, providing a picture of the health of the countryside. You can join in its 'Common Plant Survey' by simply counting primroses or other common species in measured plots of land.

RSPB

(www.rspb.co.uk)
The Royal Society for the Protection of Birds works throughout the UK. One of its aims is to stop the recent decline in songbird populations. There are hundreds of vacancies for volunteers all over the country. From becoming a local group leader to quietly surveying birds on a nature reserve, you can make a contribution in many ways.

Trees for Life

(www.treesforlife.org.uk)
This Scottish charity is dedicated to the regeneration and restoration of Caledonian forest in the Scottish Highlands. If you fancy a week in glorious Glen Affric carrying out practical conservation work, then sign up for one of its volunteer work weeks. You'll be doing anything from tree planting to collecting Scots pine cones, all under the watchful eye of a group leader.

Vincent Wildlife Trust

(www.vwt.org.uk)
For more than 25 years this charity has been supporting wildlife conservation through research and the management of reserves. It focuses on mammals, especially bats, polecats, pine martens and dormice. If you happen to spot a pine marten, give the trust a ring to report your sighting.

Wildfowl and Wetlands Trust

(www.wwt.org.uk)
The WWT runs nine visitor centres in the UK, giving members of the public a close look at spectacular wetland birds. If you want to sponsor a swan or adopt a goose, take a look at their website. If you'd like to get actively involved, you can help collect information for Britain's first 'Winter River Bird Survey' by counting water birds on a stretch of river.

Wildlife Trusts

(www.wildlifetrusts.org)
This is a network of 47 local Wildlife Trusts and Wildlife Watch, the junior branch, which between them cover every corner of the UK. They work to protect wildlife in both town and countryside, and they care for more than 2560 nature reserves. They rely on a huge network of volunteers. If you'd like to help, contact your local branch. There's something for everyone, from spending an evening counting bats to getting mucky restoring a river.

Woodland Trust

(www.woodland-trust.org.uk)
This charity protects our native woodland. There are a variety of ways to get involved, from becoming a wood warden to giving talks to schoolchildren. Or you can put your wellies on and plant some trees. See the website to find out about events in your area.

CHAPTER SEVEN

The Future

What the future holds is mainly up to us. Nature is always changing – species come and go, the climate swings from warm to cold, and sea levels rise and fall. But our activities can profoundly affect these natural patterns. There are some excellent centres that put this in context and offer advice on what you can do to help safeguard our future.

1 Eden Project, Cornwall

GRID REF SX048547

Opened in 2000 and built in the remains of an old Cornish china clay pit, this unique facility is home to over 100,000 plants, representing 5000 species from all over the world. Two gigantic conservatories form the centrepiece. The Humid Tropics Biome, the world's largest greenhouse, is filled with rain-forest plants, whilst the warm Temperate Biome contains plants from the Mediterranean region as well as South Africa and California. The goal of the Eden Project is to explore all aspects of ecology and our interaction with the planet.

Further information:
www.edenprojet.com

2 Wildwalk@Bristol

GRID REF. ST583725

This state-of-the-art centre in the heart of Bristol takes you on an amazing journey through the natural world. Over 95 per cent of all described animal species are smaller than a hen's egg and Wildwalk brings their world to life in vivid close-up, using the latest in sound and vision technology. There's also a walk-through tropical rainforest complete with free-flying birds and butterflies and the centre is home to the 'ARKive' library – the world's first-ever digital library of photos, sound and film of endangered species.

Further information:
www.at-bristol.org.uk/wildwalk

3 Centre for Alternative Technology, Powys

GRID REF. SH756043

CAT is an environmental charity which aims to 'inspire, inform and enable people to live more sustainably'. Practical displays show the latest developments in renewable energy (free power from the sun, rain and wind), environmental building design, energy efficiency, recycling and organic growing. CAT is committed to showing how people, nature and technology can live together successfully.

Further information:
www.cat.org.uk

4 Royal Botanic Gardens, Kew

GRID REF. TQ184767

Founded in the eighteenth century, Kew is one of the great botanical gardens of the world. Covering over 300 acres (230 hectares), the gardens contain over 40,000 plant varieties and 39 listed buildings, including the magnificent Palm House and Temperate House and the famous pagoda. But Kew is much more than a garden – the goal is to 'enable better management of the Earth's environment by increasing knowledge and understanding of the plant and fungal kingdoms'. Kew was awarded World Heritage status in 2003 in recognition of its historic landscapes and outstanding buildings and its important role in science and plant research.

Further information:
www.rbgkew.org.uk

EDEN PROJECT

Bibliography

1

Bennison, G.M. and Wright, A.E.
The Geological History of the British Isles
Edward Arnold, 1969
ISBN 0713122064

Fortey, Richard
The Hidden Landscape
Jonathan Cape, 1993
ISBN 0224036513

The Geological History of the British Isles
Open University Press, 2001
ISBN 0749235713

Toghill, Peter
The Geology of Britain
Airlife Publishing, 2002
ISBN 1840374047

Wood, Robert Muir (ed.)
On the Rocks
BBC Further Education Advisory Council, 1978
ISBN 0563162112

2

Andersen, Bjorn G. and Borns, Harold W. (eds)
The Ice Age World
Aschehoug AS, 1997
ISBN 8200376834

Barry, R.G. and Chorley, R.J.
Atmosphere, Weather and Climate
Routledge, 1998
ISBN 0415160200

Barton, M. et al
Wild New World
BBC Books, 2002
ISBN 0563534257

Bolles, Edmund Blair
The Ice Finders: How a Poet, a Professor and a Politician Discovered the Ice Age
Counterpoint Press, 1999
ISBN 1582430306

Gribbin, J. and Gribbin, M.
Ice Age: How a Change of Climate Made Us Human
Penguin, 2003
ISBN 0141007303

Hancock, Graham
Underworld: Flooded Kingdoms of the Ice Age
Penguin, 2003
ISBN 0141000171

Wilson, R.C.L., Drury, S.A. and Chapman, J.L.
The Great Ice Age: Climate Change and Life
Routledge, 2001
ISBN 0415198429

3

Crawford, Peter
The Living Isles
Guild Publishing, 1985
ISBN 0563203692

Ingrouille, Martin
The Historical Ecology of British Flora
Kluwer Academic Press, 1995
ISBN 0412561506

Miles, Archie
Silva: The Tree in Britain
Ebury Press, 1999
ISBN 0091867886

Pielou, E.C.
After the Ice: The Return of Life to Glaciated North America
University of Chicago Press, 1991
ISBN 0226668126

Rackham, Oliver
The Illustrated History of the Countryside
Weidenfeld & Nicolson, 1994
ISBN 184188104X

Trees and Woodland in the British Landscape
Weidenfeld & Nicolson, 1995
ISBN 1842124692

Darvill, T.
Prehistoric Britain
Routledge, 1987
ISBN 0713451807

Davies, J.
The Making of Wales
Sutton Publishing, 1996
ISBN 0750911425

Hall, S. and Haywood, J. (eds)
Penguin Atlas of British and Irish History
Penguin, 2001
ISBN 0140295186

Hoskins, W.G. and Taylor, C.
The Making of the English Landscape
Hodder & Stoughton, 1988
ISBN 0340399716

Muir, R.
Lost Villages of Britain
Michael Joseph, 1982
ISBN 0718120361

Villages of England
Thames & Hudson, 1992
ISBN 0500015295

Simmons, I.G.
An Environmental History of Great Britain
Edinburgh University Press, 2001
ISBN 0748612831

5

Aalen, F.H.A., Whelan, K. and Stout, M.
Atlas of the Irish Rural Landscape
Cork University Press, 1997
ISBN 1859180957

Hoskins, W.G. and Taylor, C.
The Making of the English Landscape
Hodder & Stoughton, 1988
ISBN 0340399716

Muir, Richard
The New Reading of the Landscape
University of Exeter Press, 2000
ISBN 0859895807

Percival, John
The Great Famine: Ireland's Potato Famine
BBC, 1995
ISBN 0563371323

Trinder, Barrie
The Making of the Industrial Landscape
Weidenfeld & Nicolson, 1997
ISBN 0753802686

6

Harvey, Graham
The Killing of the Countryside
Random House, 1997
ISBN 0224044443

Oliver Rackham
The Illustrated History of the Countryside
Weidenfeld & Nicolson, 1997
184188104X

Simmons, I.G.
An Environmental History of Great Britain
Edinburgh University Press, 2001
ISBN 0748612831

7

Currie, Ian
Frosts, Freezes and Fairs: Chronicles of the Frozen Thames and Harsh Winters in Britain Since AD 1000
Frosted Earth, 1996
ISBN 0951671081

Godrej, Dinyar
The No-nonsense Guide to Climate Change
Verso Books, 2001
ISBN 1859843352

Hunter, Robert
Thermageddon: Countdown to 2030
Arcade Publishing, 2003
ISBN 1559706678

Lynas, Mark
High Tide: News from a Warming World
Flamingo, 2004
ISBN 000713939X

Motavalli, J.
Feeling the Heat
Routledge, 2004
ISBN 0415946565

UK Climate Impacts Programme (UKCIP)
Building Knowledge for a Changing Climate: The Impacts of Climate Change on the Built Environment
UKCIP, 2003
Available to download from www.ukcip.org.uk

Weart, Spencer R.
The Discovery of Global Warming
Harvard University Press, 2003
ISBN 0674011570

Acknowledgements

It would have been impossible for me to have produced this book alone, so broad is its compass and so diverse its subject. For this reason I am indebted to all the researchers and producers of the *British Isles – A Natural History* television series for their skills, their knowledge and their enthusiasm in piecing together the story of our amazing islands, and for making the job of telling the story such a fulfilling one. In particular, I owe the following my heartfelt thanks for their contributions to this book: Stuart Armstrong, Chris Cole, Mary Colwell, Ian Gray, Patrick Morris, Jessica Pailthorpe, Charlotte Scott, Venetia Scott and Dan Tapster.

To the Open University academics and scientific advisors who kindly read the chapters and made helpful comments – Patricia Ash, Chris Bissell, David Gowing, Ian Simmons, Christopher Taylor, Barrie Trinder and Chris Wilson – go my thanks for their painstaking trouble and effort.

The BBC's Natural History Unit has a reputation second to none. The prospect of working with them was daunting, but the reality turned out to be both rewarding and tremendously exhilarating. I owe much to Mike Gunton, the programme's Executive Producer; Michael Bright, the Supervisory Producer; and Neil Nightingale and Patrick Morris for having had the idea in the first place.

To cameramen John Brown, Robin Cox, Brian McDairmant, Chris Openshaw, Tim Shepherd and Gavin Thurston, sound recordists Jake Drake Brockman, Graham Ross and Bill Rudolph I am grateful for making me look human and sound reasonably intelligent. Thanks also to off-line editors Tim Coope, Chris Mallett, Jo Payne, David Pearce and Vincent Wright and to Dave Corfield at 422 Ltd for the programme's excellent graphics.

My final thanks go to the core production team, the men and women whom I met at the crack of dawn and whose contributions made the series so special: Stuart Armstrong, Ben Aviss, Lesley Bishop, Mary Colwell, Dave Cox, Tom Clarke, Chris Cole, Andrew Graham-Brown, Ian Gray, Anna Kington, Jessica Pailthorpe, Charlotte Scott, Miranda Sturgess, Dan Tapster, Liz Toogood and Catherine Worrall. Together we climbed mountains, scaled glaciers, journeyed to remote islands and bird-infested rocks, flew helicopters and jet planes, were cut off by the tide, delighted by everything from red squirrels to Manx shearwaters, grey seals to swallowtail butterflies, ate and slept badly, worked long days, were soaked by rain and hail, burned by sun and salt spray over the space of a year and a half, and still came out of it not only speaking to each other, but sharing a laugh or two as well.

For all of us, the making of the series and the production of this book have been unforgettable for a variety of reasons. It has also reinforced our belief that there is nowhere on Earth half so astonishing as our small group of islands.

A.T.

Picture Credits

BBC Books would like to thank the following individuals and organizations for providing photographs and for permission to reproduce copyright material. While every effort has been made to trace and acknowledge copyright holders, we would like to apologize should there be any errors or omissions.

Alamy 30, Leslie Garland Picture Library 62–3, 71r; **Heather Angel** 49; **Ardea** 10br, 12, 14m, 43t, 59, Adrian Warren 6bl, 14t, 17, Richard Vaughan 48br, Bob Gibbons 90–1, John Daniels 95b, P. Morris 112–13, Richard Waller 128, Geoff Trinder 138b, John Mason 151; **Beamish Open Air Museum** 146; **BBC** Chris Cole 2–3, 51, 104–5, Mary Colwell 6–7, Charlotte Scott 119b, 135; **British Geological Survey** 31; **British Museum** 92; **Neil Clarke** 11b, 32t; **Bruce Coleman** 6br, 7b, 138t, 165b, 172, Mary Owell 42, Anders Blomquist 38bl, 43b, Fred Bruemmer 84b, Christer Fredrickson 91b, Pacific Stock 143b, 152b, Kim Taylor 165t, Chris Gomersall 176–7; **Collections** 195, Michael St Maursheil 34, Richard Gould 56–7, Dorothy Burne 84t, John D. Beldom 117, 207, Gena Davies 122, Ed Gabriel 118bl, 125, Sandra Lousada 140–1, Steve Benbow 161, Clive Shenton 203; **Corbis** 35, 103, Sandra Vannini 101; **Empics** 93, 166br, 187; **Mary Evans** 136; **Martin Farr** 53t; **Paul Felix** 109; **FLPA** Tony Hamblin 188–9b, 209; **Forestry Commission Picture Library** 65; **Garden Matters** 76t, 76b, 90br, 106t, 106b; **Getty Images** 144; **Holt Studios** 142–3; **Robert Harding** 1, 10–11, 13, 23, 27, 36–7, 39b, 46–7, 54, 73, 95t, 108, 111, 115m, 126–7, 132–3, 142br, 158, 188–9t, 191b, 200; **Hulton Archive** 41, 126b, 145, 148, 149, 185; **Rodger Jones Fossils.eu.com** 21b; **Andrew Lawson** 164; **Museum of Fine Arts Boston 2004/Bridgeman Art Library** 131; **Chris Musson Clywd-Powys Archaeological Trust** 123; **National Trust** 134r; **Natural History Museum** 10bl, 33, 118br; **Nature Photographers** 180; **Nature Picture Library** 64, Alan James 62bl, David Kjaer 96b, Dave Watts 127m, William Osbourne 127b, Niall Benvie 147; **Norsk Polarustitutt/Greenpeace** 182; **Oxford Scientific Film** 40, 45, 76m, 77t, 81, 96–7, 98–9, 127t, 142bl, 150b, 160, 166bl, 167b, 179t, 183, Mark Hamblin 21t, 150t, 153, S. Packwood 32b, Lon E. Lauber 48t, Philippe Henry 63b, 78l, Tom Ulrich 70, 72l, Niall Benvie 71l, Norbert Rosing 72r, Terry Heathcote 78r, Michael Dick 79t, David Cayless 94, Konrad Wothe 100b, Daniel Valla 110t, Mike Powles 120, Deni Brown 134t, Gordon Maclean 134l, Chris Knights 138m, 157, Michael Leach 152t, Raymond Blythe 170, K. G. Vock 173t, 179r, Mary Plague 178; **Rex Features** 181; **John S. Robinson** LRPS 24–5; **RSPB** 110b; **Science and Society Picture Library** 121; **SPL** 168; **Thoroughbred Photography** 186; **Topham Picturepoint** 5, 190–1; **Ulster Museum Picture Library** 79b, 129; **Welsh Slate Museum** 204; **Woodfall Wild Images** 68–9, 88–9, 90bl, 115t, 115b, 152m, Martin Smith 4, 100t, Jeremy Moore 14b, D. Woodfall 38br, 48bl, 52, 60–1, 67, 80, 82, 86–7, 102, 118–119, 166–7, 174, 196–7, 199, 205, Tom Murphy 38–9, Steve Austin 62 br, 192–3, Bob Gibbons 74–5, 77b, 107, Maurizio Biancare 78m, P. L. A. MacDonald 137, Mark Hamblin 154–5, 163, Ashley Cooper 156, Val Corbett 169, Bob Glover 173b, Mike Lane 179b, Nigel Hicks 197, E. A. Janes 201.

All illustrations by Jerry Fowler, except the diagrams on pp. 18–19 by Gary Hincks and the map on p. 58 by Olive Pearson.

Index

Page references in *italics* refer to illustrations